CARING FOR COWS

SALLY.

Valerie Porter

CARING FOR

COWS

with illustrations by Sally Seymour

WHITTET BOOKS

First published 1991

Text © 1991 by Valerie Porter
Illustrations © 1991 by Sally Seymour

Whittet Books Ltd, 18 Anley Road, London W14 0BY

Design by Paul Saunders

All photographs are from the author's collection, except the following;
the author and publisher are very grateful for permission to reproduce:
Anna Oakford, p. 34 (middle), pp. 74-5 (all), p. 79;
Sue King, p. 30 (below); Jane Paynter, p. 30;
Rare Breeds Survival Trust, p. 33 (bottom);
Milk Marketing Board, p. 34 (bottom).

British Library Cataloguing in Publication Data

Porter, Val, *1942–*
    Caring for cows.
    I. Title
    636.2083

    ISBN 0905483936
    ISBN 0905483944 pbk

Typeset by Litho Link Limited, Welshpool, Powys, Wales
Printed and bound by Bath Press

Contents

Introduction

CONSIDERING COWS

Picture this. A city in the rush-hour — noise, heat, colour, dust, bustle and traffic. A lop-eared, humped white cow with dark, oblique, peaceful eyes wanders inviolate and chooses her resting place casually. With a gentle, satisfied sigh she lowers herself to the ground, front knees first, settling down comfortably to chew her cud and to doze.

She has chosen to rest bang in the middle of a major road junction and the traffic is brought to a screeching, shuddering, hands-and-eyes-to-heaven halt. In no time at all it has backed up for five miles in every direction and the entire city of Delhi has been brought to a standstill by one obviously contented cow.

Contentment is the essence of cows and it is contagious, infecting those who care for them, especially if they are allowed to live at their own slow pace and are treated with consideration. They offer excellent therapy to the neurotic, the restless, the sad, the distressed and the lonely, and they also have the drollness and gentleness to make very good company. Mind you, they can also be stubborn, perverse and provocative, like any other creature, but that is usually because they have been hurried, baulked or misunderstood. One of my aims is to help you to understand them so that you are better able to care for them as they deserve.

They are also thoroughly practical, down-to-earth animals who happen to be most generous with their produce, which is why they have been exploited all over the world. This book is not about exploiting the cow, but is for people who like cows enough to regard them as allies and friends rather than commodities but who also want to supply dairy produce for the house and perhaps raise a calf or two for the freezer so that the cow covers the cost of her own keep and at least breaks even, if it does not make a profit for her owner. Such a cow, placed under no stress to maximize her yields and depending on her breed and character, will probably flood you with about 3,000 litres of milk a year, including 20 litres a day during the peak of her lactation, of which only a few litres a day can be suckled by her own calf — and that still leaves a lot of milk for the house.

The major themes of this book are *welfare* (kind caring based on understanding natural behaviour) and *good husbandry*, based on organic principles. The emphasis throughout is on 'thinking cow' and acting in the cow's best interests rather than your own, though in fact there is no conflict: a contented cow is a productive cow, and that means a contented cow-keeper.

THE ORGANIC ATTITUDE

The welfare of the animal is at the heart of any organic farming system which embraces livestock. But what *is* organic farming? Organic farmers (or gardeners, for that matter) seek to work *with* rather than *against* nature.

Organic farmers seek to avoid damaging the environment in general and the structure of the soil in particular. The welfare of the soil is the fundamental basis of organic farming, including livestock farming; any profits from its productivity should be a bonus rather than an overriding aim. Economists might smile wryly and mutter about impractical naivety but organic methods, viewed in the wider perspective of the health of the planet and its occupants, are in fact economical. They aim to avoid adding to society's increasing problems with the pollution of land, water, air and food. They do not rely on finite energy resources from fossil fuels but aim to recycle, within the holding, energy from renewable resources.

For the livestock farmer, on whatever scale, the principle of welfare is naturally extended to animals as well as the soil and, at the same time, it is appreciated that the animals themselves make important contributions to the health of the soil and that it is a mutual exchange. As applied specifically to livestock, the main organic husbandry principles are:

- Very high welfare standards, including giving the animal every opportunity to express the fullest range of its species' normal behaviour patterns.
- Maintenance of good health by preventive husbandry, good welfare practices, and appropriate housing and feeding.

- Feeding in a way suited to the animal's physiology, using home-produced food as far as possible.
- Avoidance of routine prophylactic drugs as used in conventional veterinary practice.

These organic principles are integral to the philosophy of this book and feature in more detail in appropriate chapters.

COWS FOR CHOICE?

The most successful cow-keepers look after cows because they like the beasts — and that is as good a reason as any for keeping them. It is also essential, whatever the aims.

In practical terms the choice of livestock includes ruminants (cows, sheep, goats, lamoids, deer), monogast grazers (horses, mules, donkeys, rabbits), omnivores (pigs) and poultry. Grazers are able to convert food which humans find inedible (grasses) into products which humans find useful such as meat, milk, manure, fibre and muscle power. If you have grass, ruminants are its most efficient converters, but your decision about which species of ruminant should be based on practicalities as well as emotional preferences. If you are wondering whether cows would be appropriate for you, consider these factors in comparing cows with other ruminants.

Cows are big. A factor which has disadvantages and advantages.

Cons	Pros
• They need *space*. Depending on the land, allow at least an acre a cow for grazing and another for making hay.	• They produce more *milk* and *meat* than other types of livestock.
• They need *bulky food*, and more of it per unit of production.	• They are not fussy feeders and make use of a very wide range of feedstuffs, including rough grazing and agricultural by-products.
• They need lots and lots of fresh *drinking-water*.	

Cons	Pros
• They are *heavy* enough to push through flimsy barriers and they hurt when they accidentally tread on your foot.	• Their bulk is useful for *work*: they are slow, patient, dependable, and easy to train to agricultural work or to pulling carts.
• They cannot be *picked up* and carried.	• They don't jump about like goats, nor get worried by dogs like sheep.
• They are a relatively expensive *investment*.	• They bring in relatively higher *returns*.
• They *poach* wet fields.	• They *manure* them too!

Cows are hardy. They can live out all year round, given access to minimal shelter, but they will turn a wet winter field on heavy land into a quagmire. Not that they mind too much: some of them delight in wallowing as long as they also have access to dry ground. If the field cannot take the pressure, their winter housing can be very basic and cheap: dry back, dry bed is all they need. They tolerate cold much better than heat. They are not labour-intensive; they can basically fend for themselves given adequate pasture-ranging and shelter, though, like all livestock, they should be checked *at least* once a day in case of unexpected problems. They are intrinsically healthy (unless under stress in a commercial set-up) and don't suffer from all those unpleasant parasites and diseases sheep seem to be prone to, nor do they need cossetting from the weather like goats, nor do they strangle themselves on their own tethers.

Cows must calve before they give milk. This very basic fact is sometimes not appreciated: a cow does not make milk until she has given birth to a calf, and she tends to reduce her milk supply gradually over the next few months as the calf switches increasingly to solid food. Most dairy cows are deliberately dried off and given a rest from milking about 6-8 weeks before giving birth again and starting a new lactation. It is possible to keep a cow in milk for more than a year, but unusual, and the quality deteriorates. When a cow is in milk, if she is not suckling a calf, she *must* be milked regularly *every day* without fail, usually twice a day. That is a very big tie for the milker.

Cows are easy to breed. They usually have one calf a year —
multiple births are rare, and calving problems are also rare: you
don't usually need to be in attendance. The pregnancy is nine
months — much longer than for smaller species — and a heifer
(a young female) should not be put into calf until she is properly
grown, i.e. at least 18 months old (though she might be sexually
mature several months younger). Thus it takes a long time to
build up a herd or see the results of selective breeding. But cows
can have long, productive lives; a respectfully treated cow not
put under undue stress and productive pressure could go on
having a calf a year for twenty years or so, with luck. And a
good-natured cow will happily rear a succession of foster calves
for you — perhaps as many as six or eight a year . . . What, by
the way, will you do with the calves as they grow up — keep
them, sell them or eat them? It can be a dilemma for some
people, especially with male calves or if you only have limited
grazing space.

You do not need a bull on the holding to breed from a cow. It
is very easy indeed to have her artificially inseminated — just a
'phone-call away. That gives two big advantages: you do not
have to accommodate a bull (which can be very expensive and he
also needs expert handling) or transport your cow to one
elsewhere, and you have a very wide choice of sires from all over
the country and of every conceivable breed, their semen not only
brought to your holding but also inseminated by an expert for a
few pounds. So it is easy to build up a well balanced genetic
spread in your herd from bulls whose qualities have been
thoroughly tested and proven.

Cows are generous. They give a great deal of milk (the flood
can be overwhelming unless you convert it); they give you calves
for meat; they give you a great deal of manure in big, loose
dollops rather than tidy pellets like goats and sheep or firm
scoop-uppables like horses. They dung and urinate copiously if
upset. They are also steady, slow workers: they can draw a
plough, pull a cart, carry a pack or be ridden (uncomfortably).
At the end of their life they give you a hide for leather, but they
do not give you an annual fibre crop like sheep and hair-goats.

Cows are affable most of the time. They like a regular routine
and, in return, are easy to manage but, like any other farm
animal, they don't like the unfamiliar and are readily upset by
human violence (verbal or physical). They are happy to share a
field with other livestock, especially sheep, which makes for

good husbandry. They are sociable animals and hate to be alone except when calving or sick: they *must* have company in the field, preferably of their own kind. They can be very good companions to familiar humans and are lovely and warm to lean against . . .

Cows have voices but they are not often heard. However, a cow who is 'bulling' (ready to mate) will shout her head off for hours on end for a couple of days unless mated, and a cow deprived of her calf will bellow heartbreakingly for two or three days (and nights). A lone cow will call for you if you are late for milking or feeding, but not persistently; a group of cows will call enviously if one of them has escaped through the hedge, and the escapee will call in an attempt to maintain contact with the herd. A cow with a calf has a very special mothering tone which is delightful to hear and very intimate.

1. Cow Habits and Handling

Cows can inspire great affection; they can return it, too. Their company can be rewarding in itself, quite apart from the bonus of their produce.

It is only when you get to know cows as individuals that you will appreciate them and here the smallholder has a huge advantage over the commercial cow-keeper. Cows are characters — each one of them — but sometimes their individuality gets lost in a large herd and is further muffled by them being tagged with numbers rather than names. Two one seven can hardly compete with Gutso or Fidget as a descriptive tag. I have no idea why so many cows are named after flowers; nicknames usually suit them better.

The art of keeping contented cows is to know them, as a species first and then as individuals within the species. Before acquiring your own, spend many long hours studying herd behaviour in someone else's fields and yards; and when you do get your own, spend plenty of idle time with them, watching, learning, just being with them so that you begin to know them and also let them know you. You've broken the barriers the day you can wander into the field unnoticed — when they don't move away but they don't come rushing up expecting food either. They simply accept you and allow you to be in their company. That moment comes even sooner if you are hand-milking them twice a day and make a point of lingering over the whole business, finding time to chat and to groom them and scratch their favourite places.

THE SOCIAL COW

A cow is sociable. She is a herd animal, she needs company, and is deeply disturbed by isolation. She likes the security of being in a small, familiar group with a recognized hierarchy and it is almost as unfair to keep one cow on her own as it is to force her into a large, anonymous commercial herd. She needs to be able to express her natural behaviour in interaction with other cattle. Humans, however well intentioned, are poor substitutes for the herd: we are not much good at mutual grooming and don't have rough tongues, nor do we offer a mutual tail-switch to ward off flies in a summer meadow. Nor are we long enough in the field or cowshed to become part of the herd. We keep going away again, leaving the house-cow on her own.

Wild cattle tend to live in groups of perhaps 15 to 20 cows with their calves and youngsters. Males form separate bachelor groups, leaving the cow herd when they are perhaps a year or eighteen months old but keeping within range and ready to answer the demands of a cow on heat in due course. Within the group, whether cows or bachelor bulls, the members establish a more or less linear social ranking and each small cow herd has a dominant animal at the top of the ladder, or sometimes a balanced triumvirate. There is a certain amount of jostling for better ranking but no more than general body language to assert superiority, with the occasional reminder of a bit of leaning, barging or casual butting to reinforce a cow's position. (Bulls are much more aggressive about the whole ranking business and become strongly territorial as they mature.)

Left to itself, a cow herd sorts out its **pecking order** (known as 'hooking' order for cows with horns or 'bunting' order for those without) and remains a stable group for a long time. However, the balance can be severely disrupted by the sudden introduction of strange animals or the arbitrary division of the herd for management purposes by humans; the accepted bunting order is thrown into confusion as every cow seeks to establish new relationships and it takes a lot of jostling to sort it all out again to the general satisfaction.

Commercial herds are often far larger than the typical wild herd and this in itself puts added stress on to the cows. They tend to form smaller groups within the herd, and it can be quite traumatic for the group to be split up or forcibly mixed with newcomers. It is common practice in many large dairy herds to separate out cows at different stages in their cycles for the sake of easier management; in particular, those that are 'dry' (no longer giving milk) for a couple of months before calving are kept apart from the rest of the herd until they have calved, and then reintroduced to join the milking-parlour routine after an absence of perhaps two months. They don't like it at all; they have probably been separated from their particular mates in the first place, and then have to re-establish their rank when they rejoin, and that on top of the inevitable stress of giving birth and having the calf removed, which is standard practice in commercial herds. Talk about postnatal depression!

The ideal situation, from a cow's point of view (and that is the angle of this book), is to remain in a stable group all her life, from calfhood to old age; its only new members would be growing calves and its only separations would be by death.

Within the ranks, the dominant cow is not usually the group leader. Dominance gives her prime access to the best of everything but there is usually a middle-ranking animal who is always the first to do anything — the first to head for a drink or the shade, the first to settle down for a rest, the first to get up again, and the first to wander off in search of fresh grazing or towards the milking area. The rest of the group, in their usual unhurried way, follow her example so that they act as a herd, albeit not instantly. If the leader's action results in a general move, they amble after her more or less in a line, with the dominant cow, sure of her superiority, somewhere near the middle of the order but with the lowest rankers always at the tail end.

Without human interference, cows establish their own daily routines, alternating periods of grazing with periods of loafing and resting. The **major grazing periods** seem to be at daybreak, mid-morning, mid-afternoon, dusk and midnight, more or less. In between concentrated grazing, they probably spend from about nine to twelve hours *resting*, during which they chew the cud and generally ruminate in all senses of the word, either standing or lying down. They usually lie on their sternums, forelegs tucked under them, one hindleg beneath the body. Occasionally a cow lies stretched out on her side but only for a short while. They do not actually sleep very much — perhaps about three or four hours in twenty-four — when they lie in the usual position but with the head turned back towards the flank. Rest is essential and adequate sleep is important: a cow under stress will not sleep well and you will notice a tell-tale drop in her milk yield.

Cows are creatures of habit. They like familiar routines and they are wary of the unfamiliar, though sometimes hesitant curiosity overcomes suspicion — the more readily if food is involved.

They are *slow and deliberate* by nature. Let them be so. Keep your own movements slow and deliberate too. Don't hurry or hassle them: the more you rush them, the more perverse they become.

COW SENSES

A cow's keenest sense is that of *smell*, which she uses constantly in selecting her grazing. It is also an important sense in her social life within the herd and in addition is often an early-warning system at the approach of a predator. Her sense of *hearing* is good, too — look at those large ear flaps, which can turn independently to help her judge the source of a sound.

Cows are *prey*. Unfamiliar humans and dogs are potential predators, to be avoided or, in emergencies, to be chased off by the group. Like those of most prey animals, a cow's eyes are placed at the side of the head and her *eyesight* is good for panoramic vision; she has a wide peripheral range and can sense a predatorial approach out of the corner of the eye, as they say. She tends to get a general impression rather than a detailed image. The eye protrudes, more so in breeds like the Jersey, and approaching objects loom up on her rather suddenly. For that reason, if you stretch out your hand towards a cow too suddenly it will startle her considerably. Cows are partially *colourblind* and the colour they are least likely to distinguish is red.

A cow in a large herd can probably *recognize* up to sixty or seventy individuals, using a combination of vision to recognize by stance and body language at a distance, and above all smell. Cows recognize individual humans in the same way and they will remember a familiar person even after several years' absence, especially if you have been particularly friendly or unpleasant. You won't be greeted with a doglike display of enthusiasm but you will be quietly accepted and made to feel that you've never been away. Very reassuring! A cow also has a very long memory for a bad experience, associating it with a place or person as appropriate, and there have been tales of cows exacting revenge after several years . . .

Cows are *wary* of:
- Insecure footing: slippery surfaces, sudden changes of texture underfoot, stripes or channels across their path, hollow-sounding or moving surfaces like ramps or bridges.
- Unfamiliar or steep slopes: their instinct is to go up rather than down but preferably neither.
- Shadows and entering dark places.
- Flapping clothes, sudden movements on their visual periphery.
- Sudden rustling, and the buzz of gadflies.
- Breaks in a familiar routine or subtle changes in your behaviour (don't even *think* it . . .)

COWS NEED SPACE

Every animal, large or small, deserves space appropriate to its needs rather than the space we might grudgingly provide. Cows need:

Home territory — a familiar area which is their basis of security and somewhere to which they can return at will.

Range — they are natural nomads, moving about in search of pastures new when the need for fresh grazing dictates or simply when the urge takes them (or their group leader); they enjoy rambling for the hell of it and appreciate variety.

Flight space — the distance maintained by an animal between itself and a potential threat. This is an important factor in cow-handling: the better she knows you, the closer you can come without distressing her, but she will want to keep at a safe distance from strangers and will be under stress if unable to move away from them.

Head space — this is a combination of social space and flight-or-fight space. The area around a cow's head is the most jealously guarded personal space and you could provoke an automatic defensive reaction if you unexpectedly intrude upon it — a quick hook of the horns or a head butt. In cattle, this space is about a metre in radius. Remember that peripheral vision . . .

Cows are not *aggressive* but in some circumstances they will be aggressively defensive, especially against an apparent predator and especially if there is a calf to protect. They often act as a group to chase off a dog.

Most cows are remarkably nice natured. A very few are bad-tempered, usually because of bad handling in the past or as a genetic trait (don't breed from them just in case). Their usual weapon is at the head end, whether or not they have horns, but they can also *kick*, though usually only for good reason like a worrying dog, flies tickling the udder or sore teats at milking. The action of the kick is strongest sideways and backwards.

Bulls, however, are aggressive and highly territorial. Never, never trust any bull, however docile he seems and however well you know him. Indeed the docile bull is probably the most dangerous of all, because you forget to respect him.

HANDLING COWS

Bear in mind the behaviour patterns just described and use them to draw a cow into doing what you want rather than forcing her into doing something she would rather not. Exploit her natural inclinations rather than fight them. Lure rather than drive (predators drive) and train her with kindness (and food!) so that she trusts you and is willing to be led. Use your *voice*. If a cow associates your regular shout of 'Come up!' or whatever with food, she'll probably come when you call unless she has good reason to be suspicious of your motives.

Early *halter-training* can be invaluable — starting in calfhood before the animal knows its own strength (though even a calf is surprisingly strong). Practise regularly for no particular reason, to allay suspicions when there *is* a reason!

(Left) *To persuade a cow not to fidget, hold her tail carefully nearer root than tip and nearly vertical.* (Right) *To urge a cow forward, carefully begin to curve her tail into a flat coil against her rump.*

Herding

House-cows are easy to handle as long as you treat them kindly and they can trust you. They get to know you very well indeed even if the only time you spend with them is when you milk them. Dairy herds, similarly, are the easiest to handle because they are used to close daily contact with people and they know what to expect. Dealing with a group of suckler cows on the hill is more of a problem: not only are they unlikely to have had daily physical contact with you but also they are protective of their calves.

To herd a group of younger animals not used to people, or simply full of spirit, bear in mind their distrust of strangers and remember to respect their flight space: keep out of it, or they will probably panic. Remember the herding instinct: they will bunch together as you approach. Aim to *keep* them together for easier herding. They will stay bunched as long as you remain beyond the flight space but will probably break and scatter if you intrude upon it — in which case you have lost control.

Try to stop short, keeping an eye on their behaviour (you'll sense if they are feeling too threatened), and stand calmly, letting them come forward to investigate this strange thing in their field. Keep your predatory eyes downcast and your arms hanging loose. Remember the social structure of the group; spot its natural leader and concentrate on getting that animal unwittingly on your side. If the leader goes through the gateway, the rest will probably follow.

If an animal spooks and breaks away from the group, *leave* it. Just get the rest of them where you want them to be. *Don't chase* — chasing is a predatory act and always makes matters worse. The breakaway's main aim will be to rejoin the group anyway, given the chance.

To get the group through a gateway, stand far enough away from the gap not to be a deterrent yourself. To prevent a last-minute breakaway, make yourself look a bigger side-barrier by extending your outstretched arms with leafy branches, gently shaken when necessary to move them in the right direction, away from you. Work on the *funnel* principle in any handling situation. Devise ways of gradually funnelling the animals so that they hardly notice they are going through a small gap.

Loading

Animals that know and trust you will happily come to your call or to the alluring rattle of food in a bucket but they will sense if your motive is suspicious almost before you have thought of it, especially if your aim is to confine them. Probably the most difficult task of all is to load an animal into a trailer: it is suspicious of your motives, of an unfamiliar event, of a dark place, of a ramp underfoot, of confinement and of your sense of urgency and uncertainty. Go and watch loading and unloading of batches of animals at the market or abattoir to see just how brutal people sometimes are with livestock — they get the business over and done with quickly and no doubt efficiently but their main aid is fear.

Allow *plenty* of time for loading and at no stage hurry your cow or lose your temper (easier said than done!). If she becomes agitated, stop and wait until she is calmer. Work on the funnelling principle or you will probably waste a lot of time, energy and temper chasing her all over the field.

Make the ramp as shallow as possible and make its surface less alarming by sprinkling it with sand or ashes for a firmer footing and perhaps disguising it with straw. Have *solid* side panels to funnel her up the ramp, high enough so that she is not distracted by movements and objects on either side. Try luring before driving; food or the delightful aroma of sweet hay might help. Ideally, the confined space should not look dark, and indeed an apparent exit at the other end will make her more willing to enter, even if it is only an illusion. If she is well trained, lead her in with a halter, with two other people using a rope or linked arms across her backside to heave her forward. When leading by halter, *do not let an unwilling animal get her head down*. If she does, she will soon go down herself and refuse to budge an inch.

Young *calves* are easier to manage, especially if you remain calm and kind and make allowances for their bewilderment. To carry an individual calf, put one arm around its chest and the other around its buttocks, picking it up so that its legs dangle beneath its body; do *not* wrap your arms under its belly. To lead a calf, hold it by the lower jaw.

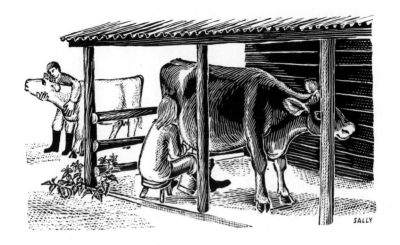

SALLY

2. Welcome

What can you offer a cow looking for a good place to live? Cows know their own minds and any sensible cow would ask the following questions before agreeing to take up residence:

- How many other cows will there be, if any?
- What are the fields like — the grazing, the exposure, the shelter, the water supply?
- What's on the menu?
- What's my job?
- Can I keep my calf?

MANAGEMENT AND WELFARE

The welfare of cattle is a matter of commonsense and good husbandry based on compassion and respect. If you want to know more about animal welfare guidelines, consult the Universities Federation for Animal Welfare (UFAW), the RSPCA or Compassion in World Farming.

The problem in British farming is that the whole business has sadly become just that — business. It has become essential, it seems, to concentrate firmly on economics, and that has meant a lesser regard for the contentment of livestock. Some people have forgotten that contented livestock are also more productive, without the need for all the additives and prophylactics that are now commonplace in commercial farming. If the basic husbandry is sound and is concerned above all with the welfare of the animals, there will be far less disease to combat; and the healthier the animal, the more it will produce and the better the quality of its produce.

You cannot teach stockmanship — it's more of a gut feeling, enhanced by clear-eyed observation and experience, and it should come *first*, before economic considerations, before playing with grants, before even thinking of acquiring animals. If you do not have the makings of a good stockman, forget it and stick to crops and tractors. It is quite painful to watch well meaning people who know all the theory, have studied at a college perhaps, and take every care to 'do it right' but who cannot recognize the tiny flickers of body language that tell the stockman an animal is unhappy long before it actually becomes ill. People scoff at anthropomorphism but there must at least be a genuine empathy and respect for the creatures in your care.

Commercial systems

A **calf** is an animal up to 180 days old.

A **heifer** is a female more than 180 days old who has not yet given birth to her first calf. A **first-calf heifer** is one who is in her first lactation. When she produces her second calf, she becomes a **cow**.

A **bullock** is a castrated bull; a **steer** is also a castrated bull, usually being reared for beef.

An **ox** is a castrated bull, usually a draught animal, but in physical fact there is no difference between bullocks, steers and oxen.

A **bull** is an entire male more than 180 days old, either for breeding or for 'bull beef'.

Store cattle are beef animals 'ticking over', being fed economically on surplus food and not expected to gain weight.

Fattening cattle are beef animals who *are* expected to gain weight and condition, and to 'finish'.

The main commercial livestock systems in Britain are dairy farming, suckler herds, artificial calf-rearing, and beef-fattening and store cattle. The first division, however, is between **intensive** and **extensive** systems. In an intensive system, high outputs are achieved by means of high inputs, and the animals are under a great deal of production stress. They give a lot (especially of milk) but they need expensive housing, feeding, equipment and labour. In an extensive system, the inputs are low and so are production levels. Typical extensive systems virtually dispense with housing, or keep it very simple, and keep labour and equipment at a minimum.

The dairy herd
British dairy herds are large — anything up to about two hundred cows — and that is their first drawback from the cow's point of view. Then there is the 'efficiency' of mechanized milking (which distances the milker from the living animal), concentrate feeding (probably the major cause of BSE — see Chapter 5) and winter housing systems like cow cubicles, designed to minimize human labour by giving the cow a concrete bed. The result of these management systems can be poor health.

On top of all that, the cow must face the trauma of losing her calf. In a commercial herd calves are removed from their mothers within three or four days of birth, permanently. This is essential as the whole point of the herd is to have its milk extracted for human consumption, not for calves. A cow's milk for the first three to four days after she has calved is thick and yellow with **colostrum**, a substance vital to calves but not wanted in the bulk milk tank. As soon as the yield becomes ordinary milk it is reserved wholly for the tank. Goodbye, calf.

No doubt a cow recovers from this maternal loss after a while, though the noise she makes in shouting her heart out for the lost calf can be overwhelming. But it is an annual event — most dairy-herd cows have never in their lives reared their own calves and it does not take much imagination to appreciate the possible cumulative effects of the deprivation. It is yet another stress and the dairy cow must sometimes wonder what the hell life is all about. She is pregnant for nine months in every twelve and has nothing to show for it after all except the twice or thrice daily embrace of the milking machines in their relentless demand for *more*. She is on a production line: she is intensively fed to produce maximum milk yields, she is expected to calve every twelve months, she is ruthlessly culled if her performance flags or she fails to hold to her service (and she might never meet a bull in her life); she is even, in some herds and increasingly, subjected to the indignity of embryo transplants, developing and giving birth to a stranger's calf with which she has no genetic link and which anyway will be taken from her after birth. She is pumped full of hormones to increase a yield which is already too great for her udder and system to bear; then she is pumped full of antibiotics to counteract the problems that are the inevitable result of overproduction and crowding. She is a number on a computer print-out; she is even fed automatically by a machine which identifies her number from an electronic tag around her neck and strictly rations her to the amount it has computed as appropriate, and few herdsmen now have the time or inclination to take her personal preferences into account.

No wonder so many dairy-herd cows don't live to see their sixth birthday — which in human terms means life ends in your twenties. If you are rich, it might be a thought to set up a retirement home for overworked dairy cows . . .

The suckler herd

Cows in suckler herds have much better lives altogether. They are in extensive rather than intensive systems and they rear their own calves; they are not overfed and overworked and over-milked like pressure-cooked dairy cows. They form a stable group, remaining together for long periods, and more or less living their lives as they please. They might have to rough it out on the hills with much less lush grazing and in the teeth of the weather but that is no real problem to a cow and they are far healthier than their pampered, pressured milking cousins whose very environment encourages disease.

The suckler cow's calf, though destined to have a short life before it finds itself at the abattoir, has the luxury of living in a herd with its mother or a foster mother and learning the facts of life from her. It also benefits right from the start by acquiring, through her, a degree of natural immunity to the home farm's diseases and has the chance of being and remaining strong and healthy.

Calf-rearing

What a contrast between the life of a suckled calf and that of one reared artificially! The latter, usually the offspring of that deprived dairy cow, has been separated from its mother almost before it can recognize her as such and is unlikely ever to meet an adult cow again unless it is a heifer being raised as a replacement for the dairy herd. It has been transported from the home farm, probably quite roughly handled on the way, to an alien place — at best straight to its new home, at worst to the market en route to its new home — where it encounters a whole range of alien bugs against which it has little resistance and little willpower to fight after so many traumas in its first week of life. What a start! It then finds itself in an unfamiliar group of calves of its own age, all inexperienced and motherless and frightened, and it probably has to learn the unnatural and difficult act of drinking reconstituted powdered milk from a bucket. Deprived of the comfort of suckling, it sucks anything it can get its mouth around.

However, it gets used to the routine, shut up in its private pen and never to see an animal of a different age group, but life gradually improves as it becomes part of the group and even (oh joy!) is released into a *field* — with this strange stuff, *grass*!

Beef cattle

This is often the best part of life for artifically reared calves and not bad for suckled ones either. Very few demands are made on them, whether they are grazing or housed: their sole job is to eat, grow and put on flesh and 'condition'. They will probably go through the unpleasant ordeal of the marketplace at least once, and might be transported on journeys which could last for several hours (store cattle, for example, are often shipped across from Ireland to graze in England) and at the end they will go to the abattoir, but in between life is not so bad.

Small-scale systems

Now take a look at the possibilities on a smaller scale:

       House-cows for milk only.
       Suckler cows to rear calves (single or multiple).
       Hand-reared calves.
       Paddock-mowers.
       Working oxen.

In practice, the first two roles are often combined: it is perfectly possible to milk a house-cow (or two or more) and to let her suckle her calf as well. There is plenty of milk for both and the only problem is arguing over the teats with the calf. There are ways around this one — see Chapter 4. Such a system can be very satisfying; you get house milk as a daily return on your investment in the cow, while she has the pleasure of rearing her own calf (and some fosters, if required) so that they have each other's company — an important aspect of the whole business.

Hand-rearing calves is quite a different prospect needing quite a different attitude. It is almost as traumatic on a small scale as on a commercial one if you are buying in calves, especially from market. The odds are so heavily stacked against the little ones right from the start and it will need a great deal of tender loving care and patience to raise them successfully. You need extremely high standards of hygiene and capital investment in good accommodation. Perhaps the best arrangement is to rear a neighbouring dairy farm's heifer calves to the stage when they can be returned to the dairy herd.

Working oxen used to be a common sight in our countryside; they were the only major source of farm-power and heavy road transport right up to the 18th century, after which they were rudely displaced by the heavy horses which, in turn, ceded to 19th-century railways and 20th-century lorries and tractors. Oxen, as castrated males, grow to a considerable size for heavy, slow work, but you can also work a willing house-cow to some extent (see Chapter 6). Cattle of various kinds are used in Africa and Asia as pack animals, too, and are regularly ridden by, for example, the Baggara of North Sudan (the zebu hump gives the rider something to cling to — very necessary with such loose-skinned animals!). And Madura cattle are actually harnessed to little carts and raced, fast . . .

ACQUIRING STOCK

In principle, avoid markets, whether you are buying calves, steers or cows. Animals in the marketplace are inevitably placed under considerable stress — alien surroundings, unknown and often uncaring humans, unfamiliar cattle, loss of group and home territory, loading and travelling, and exposure to disease. There is no point in inviting trouble or importing disease into your own holding; if you do buy from the market it is essential to keep newcomers separate from home animals until you are sure that they are healthy.

Buy your animals direct from a recommended source and make a point of seeing them in their own surroundings before you do so (it is amazing how many people buy cattle 'blind'); then you can check that they come from a good background where husbandry standards are high.

Do all in your power to reduce the stress of the transfer. Use something like a horse trailer for a full-grown cow. Take care over loading and unloading. Make sure the animals have adequate food and water for a longer journey and space to lie down but not so much that they tumble over every time the vehicle goes round a corner. After unloading, let the newcomer find her feet in her new surroundings unmolested; leave her alone for a while to recover her dignity and get her bearings, with plenty of drinking-water and good hay or food. If she is to join an existing bunch of cattle, put her next to them for a few days before finally mixing her with the herd. There is bound to be a fair amount of jostling and social sorting out; let it take its course. If new animals are on their own, quietly be around so that you can start to get to know each other but wait for *them* to come to *you* when they are ready.

Choose an animal that you actually *like*, in character and in looks, as well as one that is physically appropriate to your enterprise; with luck it is going to be with you for a long time and the basis of any good relationship is a mutual liking! It goes without saying that, unless you are intent on animal rescue, choose a healthy beast.

A **healthy** cow has:
- Bright, calm eyes that are full rather than sunken or staring. They should not be runny.
- A bloom to her coat (though any cow looks a bit ragged when moulting after winter).
- Soft, pliable, smooth skin.
- Sweet-smelling breath.
- A moist muzzle, but not a runny nose.
- An easy-going interest in life in general.

Breeds

You can choose a mongrel if you wish or, more deliberately, a crossbreed of known parentage, but the point of the concept of pure breeds is that they have been developed for specific purposes and, broadly, you know what you can expect from them because certain important characteristics have been 'fixed'.

Some breeds are physically and temperamentally best as dairy cows, some as sucklers, some as beef animals or workers. Some combine their roles: those described as **dual-purpose** supply milk and meat, and those that are **triple-purpose** are also used for work. Some breeds need the relative lushness of lowland pastures to perform properly; some prefer rough upland grazing; some are tough enough to withstand Scottish island conditions of weather and sparse grazing but do very much better on the mainland.

Dairy breeds are more or less wedge-shaped, narrower at the front, whether viewed from the side or from above. The head, neck and shoulders of the cows are relatively fine, while the broadness at the rear includes a good-sized udder and wide hips that facilitate calving.

Beef breeds are more or less rectangular and blockily built. The prime cuts are at the rear (sirloin and topside) and some Continental breeds show extreme 'double-muscling' — lots of excessively lean beef but considerable calving problems for a double-muscled cow.

Draught breeds have the weight at the front end rather than the rear; they have mighty shoulders and necks.

There are more than two dozen British cattle breeds and an increasing number of imported ones but, frankly, we have the best breeds in the world so why bother with fashionable imports? There is a British breed for every circumstance here at home; they originally developed to suit specific regions in the type of environment and type of production for those regions.

Some of the native breeds are now classified as **rare**. If you are interested in preserving a heritage and are knowledgeable about breeding, it would be immensely satisfying to build up a good herd of a rare breed to help conserve it for the future. Talk to the Rare Breeds Survival Trust about the possibilities and listen very carefully to the advice of its technical adviser if you are serious about this important work. Don't play with such precious and vital assets.

The beginner's cow
Most people start with a house-cow or two to supply the family with milk, butter and cheese. The most popular house-cow breeds are probably Jersey and Guernsey (for their rich milk) and Dexter (because it is small). Those interested in larger milk yields and better beef calves might choose perhaps Ayrshire, Dairy Shorthorn or the ubiquitous black-and-whites (Friesians are chunkier than Holsteins and less likely to drown you in milk). Good breeds for suckler cows include Welsh Black, Galloway, Hereford, Highland and Devon, preferably whichever is your local breed.

Crossbreeding
There is nothing against mongrel cows as long as they suit your purpose and you like their characters and as long as you are not trying to build up a pedigree herd. Most people keep a pure-bred breeding cow and then use a bull of a different breed if they want a beefier calf from a dairy-breed cow; some use a dairy or dual-purpose breed of bull on a non-dairy cow if they want milkier heifer calves. Indeed, certain crossbreds are well known as good suckler cows, especially the Blue-Grey widely used in the Borders — a cow from a White Shorthorn father and a Black Galloway mother. Other ideas for crosses are given in the section on breeding.

The final choice
It will probably cost you a few hundred pounds to buy an adult cow but only a two-figure sum for a calf. If you are new to cow-keeping, your best choice would be an experienced cow, perhaps one whose milk yield has become too low for a commercial dairy herd or a house-cow whose owners can no longer keep her. The cow will know about calving and milking already, even if you don't. Make sure that her udder is sound (though for your purposes it probably does not matter if only three teats are functioning) and that she is already in calf — if she is not, it might be because she's barren. A good time to buy her is when she is perhaps three months before calving; the pregnancy

should be pretty safe at that stage and her milk yield has probably dropped to low enough for you to learn milking without causing her too much discomfort. However, it is advisable to know how to milk before taking on a cow in her first full flush soon after calving, so learn on other people's cows first if necessary.

Dexter cow, Knotting Little Ladyrush.

Dairy Shorthorn.

You might be tempted to buy an in-calf heifer but be wary of doing so unless you are experienced. There is no history of the heifer's ability to calve easily and she has no experience of calving, mothering or milking. You will both be learners and that could be the start of a disastrous relationship.

Another alternative is to buy calves or young heifers. Calves are much cheaper initially, but you are faced with all the costs and problems of rearing them to maturity, and, if you succeed, they are bound to be the naughtiest animals in your small herd! Young heifers will probably already be a little wild and wary and you could have considerable handling problems with them unless they can run with mature animals who can teach them a thing or two. But why make life difficult? Start with a couple of reliable old cows — and 'old' from a commercial dairy herd means they've probably got another ten years of productive life in them yet, if you treat them with consideration.

BREEDS

[* = rare breed]

ABERDEEN-ANGUS: Black, polled. Prime beef sires worldwide. Chunky, quite short but now being bred taller.

AYRSHIRE: Reddish-brown and white pied, lyre-shaped horns. Dairy. The elegant cow with the perfect udder; good yields and good quality.

*BRITISH WHITE: White with black points and freckles, polled. Dual-purpose, originally old milking breed, now beef, good for multiple suckling; quite large.

DEVON: 'Ruby' red, middle-horned. Suckler moorland cow, very hardy and sound, docile, good for beef or work, also for butter and cheese.

DEXTER: Black, dun or red, horned. House-cow, short-legged and very small to dwarf, but avoid extreme dwarf type. Friendly and adaptable, good milk yields.

FRIESIAN: Black-and-white pied, horned. Dual-purpose for milk and beef. Dutch cattle bred in Europe; similar stock bred in America as HOLSTEIN for milk, not beef — higher yields, bigger and rangier cows.

GALLOWAY: Black, also dun or red, polled. Famous uplands suckler cow, rugged beef-maker, outdoor breed, face full of character. Also WHITE GALLOWAY (white with black points) and the delightful, milkier BELTED GALLOWAY or 'Beltie' (black, dun or red with broad white belt).

*GLOUCESTER: Mahogany with white finching (line on back, over tail)' and belly, horned. Good-looking house-cow, useful for cheese; was originally also for work and beef.

GUERNSEY: Golden-and-white pied, short-horned. Dairy, rich golden milk, good house-cow, calves better beef than Jersey.

HEREFORD: Red with white face, middle-horned. Classic beef breed with worldwide influence — meat from grass. Good suckler cow, friendly and easy-going.

HIGHLAND: Various colours (creams and fawns to almost black), long 'handlebar' horns. Suckler, very rugged for uplands, very shaggy coat; long-lived, protective mothers, slow-growing calves for beef but good taste.

*IRISH MOILED: Red roan and white, polled. Very rare indeed.

JERSEY: Various colours, short-horned. Milk — the worldwide dairy lady, delightful house-cow of character, very creamy milk; small, light-footed, beautiful eyes. Coat colours mainly various shades of fawn and tan, usually darker on extremities, pale ring around muzzle.

*KERRY: Black, middle-horned. House-cow (ancestral Celtic dairy cow) but very rare. Elegant, pretty, alert, active, light-footed, hardy, good milk yields and good cheese. Touch of white on udder.

LINCOLN RED: Red, polled or horned. Very large beef breed.

*LONGHORN: Red roan with white, long-horned. Beef suckler and worker, plenty of milk too. Sweeping horns characteristic; big animals, well liked by owners, peaceful nature. Colour from dark plum brindle to light red roan, with white finch-back and thigh patch.

LUING: Various colours. Upland suckler for exposed regions; originally from Beef Shorthorn bull on Highland cows.

RED POLL: Red, polled. Truly dual-purpose, milk and beef from low-cost foods. Good cheese; udder sometimes pendulous.

Gloucester.

Longhorns.

Belted Galloways.

White Park.

British Whites.

Guernsey.

*SHETLAND: Black-and-white pied, horned. Crofter's cow, small, short-legged but bulky, very hardy and thrifty; milk and meat. Used to be various other colours.

SHORTHORN: Red, roan and white mixtures, short-horned. Classic group with worldwide influence, including the BEEF SHORTHORN (reliable suckler, quality beef, now rare at home) and the DAIRY SHORTHORN (very dependable, down-to-earth, easy-going, friendly and adaptable). WHITEBRED SHORTHORN as sire on Galloway for Blue-Grey sucklers.

SOUTH DEVON: Yellow-red, horned. Dual-purpose, large breed, once for Devon cream, now for beef. Thrives in hot climates.

SUSSEX: Dark red, middle-horned. Beef and excellent worker on heavy land; strongly built, strong legs and feet; colour almost nigger-brown. Unusually indiscriminate grazer, leaves tidy sward.

WELSH: Black (and colours), longish-horned. Hardy suckler, excellent mother, long even lactation to rear beef calves. Good house-cow too in own region. Main breed black, but many other colours here and there including belted and colour-pointed.

*WHITE PARK: White with black (or red) points; long-horned. Ancient type, pride and symbol of RBST; good beef but for specialist and experienced breeders only (too precious to waste).

HOME

A cow demands very little of her environment and is easily satisfied but the human who cares for her is less resilient. Very often, especially on commercial farms, a lot of capital and materials are invested in making livestock buildings for the convenience of humans rather than for the well-being of the cattle. Forget all that and stop to consider what the *cow* wants.

She wants options. She wants fresh air and space; she wants to feel the sun on her back and to stand in the rain but with the choice of shade or a dry back and dry bed should she wish to take it. In summer, the living is easy — out in the field with trees for shade and a simple field shelter where she can escape those aggravating flies and the heat, and perhaps a pond where she can stand knee-deep in the water if she is so inclined, with a clean tank of pure, fresh drinking-water as well. In winter her field could be a mess and she needs access to a hard surface and an open-fronted but deep building where she can take shelter and where you, too, can look after her and milk her in relative comfort. And that really is about all you need — field, tree, field shelter, water, hard surface and cow shed.

The field

This is the cow's true domain, be it on the hills or the plains. In an ideal world there would be no boundaries to the field and the cow, like the Delhi cows, could range where she pleased. In reality she must at some point be limited, which means that within those confines she should have as much choice as possible.

Cows need *variety*:
- In topography: not a flat, featureless plain but an area with hidden corners, hollows, curves, hillocks, out-of-the-way places, a thicket to loiter in, a little copse, a big tree to stand under, all providing shelter from onlookers as well as from the elements.
- In the grazing: mixed grasses with plenty of herbs, and hedgerow plants for browsing.
- In the ground itself: dry places, hard places, soft places, damp places, to give her feet different surface textures and to let her choose her own bed.

On reasonably good grazing, allow an acre of land per cow and another acre for hay-making. If she's rearing a calf, add about half an acre. A cow is not really an adventurous animal but she does enjoy watching the world go by and needs that stimulus whether she is in the field or (especially) housed during the winter. Variety brings interest but it also forms the basis of good health in the field. For example, cows are good at selecting certain plants as tonics or medicines when they feel off-colour — given the choice. Too often, fields are monocrop sheets sown entirely to one or two varieties of grass on 'weed'-free soil. There is more about field plants in Chapter 3, including those which should be eradicated.

Water — all cattle need access to ample fresh, clean water. Even a grazing cow in milk can drink up to 10 or 12 gallons a day (50-60 litres), and needs to do so. The water must be fresh. She might like wallowing in a stagnant pond but she needs running water to drink, either from an unpolluted natural spring or from a mains-connected water trough controlled by a ballcock. Her health and the quality of her milk suffer from contaminated or inadequate water.

Make sure that water remains available in freezing weather and insulate the supply pipe and the trough itself, packing it round with turf, or straw (which will be eaten) or manure (which will not, but watch out for contamination of the water). It is absolute *hell* struggling out with buckets of water for thirsty cattle in winter, or hammering at thick ice on the trough early in the morning only to find it has frozen again by the time you have finished milking. Oh, for a stream — but how do you know what pollutants are entering upstream?

Check regularly that ballcocks work properly, otherwise there will be an overflow which quickly creates a quagmire around the trough. Cattle are far too heavy for the duck-keeper's trick of laying a wire-netting carpet to reduce mud-puddling.

Poaching is a wet-weather problem: it means the trampling of a well-used area so that it becomes choppy mud rather than grass. It is the major reason for housing cows in winter if the land is heavy or badly drained — for the sake of the field more than the cow.

Boundaries

The main method of restricting a cow's ranging are hedges, walls, fences or tethering.

Hedges are excellent as long as they are well established and properly maintained so that they stay bushy from top to bottom: lay them if necessary. Hedges are more than a barrier: they also offer windbreaks and browsing. Cattle tend to lean into a hedge, reaching over it to browse or just to look around and they can find themselves on the other side almost unintentionally. You therefore need a good, thick, thorny hedge; quickthorn (hawthorn) is probably the best hedging species but mix it with other hedging plants for variety.

Drystone walls are only appropriate for regions where there are people who know how to build and maintain them, and suitable stone is readily to hand. They don't provide any browsing but they do give good shelter and secure boundaries with a long life.

Fencing ranges from expensive and handsome post-and-rails to cheap runs of electrified wire.

FENCING

Type	Spacing
Post-and-Rail	
All-wood, expensive but good-looking. Sawn or cleft timber — oak, chestnut, softwood. Mortised or nailed. Nails easily pushed out by leaning cattle. Easy to jump or shrug under.	Three rails at 20cm, 55cm, 92.5cm riven; 34cm, 69cm, 105cm sawn. Extra bottom wire needed with youngstock.
Post-and-Wire	
For straight runs. Need very strong end/corner posts to take tightly-strained wires. Chestnut posts (round and split).	
Barbed-Wire commonly used for cattle, reliable, long-lasting, but awkward to handle. Must be strong 2-ply.	Posts 2.7m apart. Strainers 50-100m. Wires at 0.7m and 1.1m, extra at 0.3m for youngstock.
Plain Wire for quiet animals but needs more strands. Mild steel (4mm) or high-tensile (3.15mm).	At least seven wires, lowest at 7.5cm above ground, top at 106cm.
Stock Netting — welded mesh, heavy weight, not for horned cattle; needs reinforcing with barbed-wire at top to prevent animals pressing it down.	Strainers up to 150m, posts 2.7m apart, top of netting 85cm, 2-3 extra wires above to 106cm.
Electric Wire for temporary use, cheap, flexible. Keep clear of grass-stems to avoid shorting.	Posts 10-12m apart. Line at 0.8-0.9m; extra lower wire for youngstock.

The old and true saying is that good fences make good neighbours and it is always worth investing in your boundaries, for the sake of your own peace of mind as well as for good neighbourly relations. Smallholders might shudder at the initial cost but it is probably cheaper in the long run to make a sensible initial investment in sound perimeter fencing than to have to panic-and-patch at regular intervals.

If your animals are all mature cows, they will not make much of an effort to go anywhere else and your fencing, though robust, need not be elaborate. However, if there are any young animals you need to be more careful; they can get under as well as over and through, and you need extra lines of defence to contain them.

Electric fencing is versatile and cheap, and ideal as a *temporary* measure — for example to partition off part of a field set aside for haymaking or to control grazing within the field — but it is not secure enough for a permanent outer boundary.

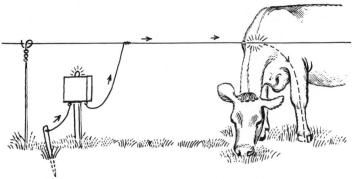

Electric fencing: The animal completes the circuit when she touches the wire, earthing the current so that she receives a mild shock.

Tethering is only a suitable option for one or two animals where the available grazing is in open spaces which cannot be fenced. It is labour-intensive and sometimes dangerous for the animal, and it is too restrictive; it does not allow a cow the freedom of expression that this book emphasizes. Don't do it unless there is no option, in which case should you have cattle at all? However, tethering does make it possible to utilize all sorts of patches of grazing — orchards, road verges, commons and headlands, for example — but you must train the cow to the tether and keep a careful eye on the animal and move it regularly to ensure that it has adequate dung-free grazing. It also needs access to water, shade and shelter, unless you are prepared to pop out at frequent intervals to see that it is comfortable and safe.

Field shelters

At its simplest and cheapest, and probably quite adequate, a field shelter can be made from very stout poles well dug into the ground to support a corrugated-iron roof set at a slope to drain off rainwater (which makes quite a din on the tin) and throw it well clear so that the area under the roof is always dry. Site it in a sheltered part of the field, protected from the prevailing wind, and preferably on higher ground for good drainage. The aim is to let the animals have dry backs and dry beds.

For further protection, especially from wind-driven rain, you can partially clad the sides. The aim is to provide weather protection rather than to confine the livestock, so the cladding does not have to be very robust or expensive. Be warned, however, that the animals will lean on it, rub against it for a good scratch and, if it is straw, playfully butt it, pull it apart and eat it so that it needs to be protected in some way. Talking of scratching, all animals appreciate somewhere for a good rub — under a low branch, perhaps, or a strong wooden rail where they can scratch backs and necks.

Cheap **cladding** materials include: old scaffolding boards, discarded pallets (scrubbed free of any chemical residues), wany-elm boards, sawmill lumber offcuts, old doors (free of lead paint), softwood thinnings (poles), wattle panels, birchbrush

'Rustic' field shelter and calf house improvised from secondhand materials.

faggots, straw bales (protected with wire-netting or wattle and very firmly secured), straw thatch stuffed between wire-netting layers, plastic windbreak sheeting.

You could invest a little more and make an all-purpose field shed to act as a ready-access shelter, calf-house, field milking parlour and place for occasional confinement (for veterinary visits), with perhaps a roofed 'verandah' attached. Properly designed and built, it could also serve for winter housing, in which case you will need to drain and pave the floor (try the local council for second-hand paving slabs or use chalk rubble etc.) and the shed's surroundings. There is plenty of scope for ingenuity, depending on your purposes, your pocket and the materials to hand. You could even buy a prefabricated field shelter or get a professional to make something elegant to your own design but it is hardly necessary in the field and anyway spoils your fun.

Cow-houses

If you need indoor housing the rules here are simplicity, ruggedness and ingenuity. The aim is to provide a dry floor under a sound roof, with partially clad walls to keep out the worst of the wind, rain and snow but admitting lots of fresh air and natural light (more important for good health and good spirits than most people appreciate). As soon as livestock are housed in too confined a space with stale air, they become susceptible to all sorts of physiological and psychological problems.

Don't stint on space. In the bad old days milking cows were literally chained to the manger for the whole winter. A Polish friend vividly remembers the first day of spring when the cows were released; they could hardly use their legs as they swayed and stumbled out of the cowshed, they were half blind after months in semi-darkness, their coats were in a state (how could they groom themselves tied up?) — and they were so excited at their new freedom that they were almost hysterical. Appalling! But even with well housed cows the spring turn-out is a joy to behold as they first feel the grass under their feet again — even the oldest cow becomes a skittish, snorting, wild-eyed, galloping, heel-kicking, udder-swinging heifer almost laughing with delight and exhilaration at the freedom . . .

The ideal winter quarters are really a field with a roof, giving plenty of room for ambling about, plenty of choice of where to lie or stand or gossip, whether to get wet or keep dry. If you have very little grass and must preserve it from being poached, then your winter quarters should have:

- Space and fresh air.
- Regular and preferably free access to a concreted outside yard for exercise, sun, air and a change of scenery.
- A dry insulated bed and a dry back.
- Water supply and food.
- A view.

In designing or converting a building, first consider its purpose and whether it is in fact necessary at all — from the animals' point of view. Many store cattle, for example, are wintered in concrete-and-straw yards, protected by concrete-block walls to keep out the wind and with perhaps a roofed area supported by poles at one end, nothing more elaborate than that. It could be even simpler, using straw bales rather than blocks for the walls, but there must be good drainage to take away excess moisture. A *pole barn*, on whatever scale, is useful to protect cattle or fodder in winter. It is no more than the simple field shelter already described but usually much larger, using secondhand telegraph poles to support the roof.

The most useful old building on a farm is probably the traditional open-fronted three-bayed *cart-shed*. I borrowed such a building to house two cows and their offspring overnight during the winter, letting them into the adjacent field during the day unless it was too muddy, in which case they had the freedom of the concrete yard in front of the shed. It was an excellent arrangement (I was very lucky to find a farmer willing to lend me the shed in exchange for house-milk) and it was also my milking parlour and a place where the animals could be confined for veterinary attention or general care when necessary. The building was deep enough for the animals to be snug at the back or to spend half the day gazing out over the front rails and sunbathing if I needed to keep them in. The interior was easily sub-divided by a system of strong poles when necessary.

You *do* need somewhere familiar where an animal can be confined for a short while under cover if it needs to be closely inspected or handled by a stranger for whatever reason, and that means not only securing its head by means of a rope halter or a neckchain (specially designed for cows) but also, preferably, side rails making a temporary pen to prevent it from moving out of the way. Everybody needs such a *holding pen* for emergencies if not for routine inspection and treatment.

Winter housing must have properly drained *floors* so that excess moisture can seep away into an appropriate effluent system. Concrete is the most versatile and probably the most hygienic floor but no cow used to the comfort of turf wants to rest on such a cold, hard floor. It needs padding, partly to cushion the cow, partly as insulation and partly to absorb dung and urine. The best such *bedding material* is straw.

You can either clean all the soiled bedding out every day or two and start afresh with new material, or simply remove the worst of the dung and pile up fresh bedding on top, letting it all accumulate during the winter and having a massive muck-out in due course. The latter practice has two main advantages: it creates a very warm bed and it begins the composting process that converts bedding and dung into valuable manure. The main drawbacks are the Herculean labour of that eventual clear-out and the fact that the surface level rises as the bedding deepens. Make sure that your barriers rise too!

The *milking area* must be kept separate, whether within the cowshed (delightfully warm on a frosty morning) or out in the fresh air. A winter-housed cow would appreciate having to walk to a separate milking area, partly for a change of scenery and also for the exercise and a chance to walk on concrete. A concrete floor is also important for parlour hygiene (it is easily sluiced down). Now all you need is winter lamplight: very romantic!

3. Feeding

Wild cattle eat what they can find, travelling as many miles as necessary to do so, but domestic animals are given little choice but to depend on us to provide them with food. Whole books, not to mention computer programmes, have been written to help farmers feed their cattle scientifically. Learn from them if you are interested in theory but, practically, you really do not need to know more than a few basic principles. The scientific angle only becomes important if the productivity demands made on the animal exceed its natural capacity.

The wild cow produces only enough milk to satisfy the changing needs of her calf — the equivalent of about a tenth of the calf's bodyweight per day during its first three weeks, after which it begins to digest solid food as well and gradually reduces its need for milk. A young calf of a Jersey size, for example, would need about 5 pints (3 litres) a day. The wild cow eats (and drinks) enough to yield that milk and keep herself fit but a modern dairy cow has been bred to fill the pail rather than the calf and even an ordinary Jersey house-cow with no pretensions will give you 5 or 6 times as much as her calf would need every day in the early weeks of her lactation. An average commercial Holstein cow produces some 6,000-7,000 litres (say, 1,500 gallons) of milk during her 305-day lactation, and the more highly bred go well over the 10,000-litre mark. Goodness knows how her udder copes with the strain and clearly she would be dead in no time without extra feeding.

Naturally a cow's stomach capacity and appetite are limited, but farmers have not let that stop them demanding higher and higher milk yields from their cows; they have found ways of giving them extra food in concentrated form so that the stomachs don't burst under the weight of it all. And that is where the trouble starts. It was concentrates, you might remember, incorporating frankly unnatural food for vegetarian cows (in the form of animal protein, and that from dubious sources anyway) that most probably caused or at least exacerbated the horrifying epidemic of BSE — bovine spongiform encephalitis.

Forget all that scientific super-feeding and take a look at the simple, basic needs of an unpressured cow such as yours.

THE DIGESTIVE SYSTEM

The cow is a ruminant, which is to say that she has four stomachs, one of which is her rumen.

The four stomachs are:
The **rumen**, taking up about 60 per cent of the space and designed to store large quantities of bulky food and begin to break down the cellulose in plant-matter.

The **reticulum**, a small pouch to which inedible objects are diverted, to be broken down or, if necessary, stored ad infinitum.

The **omasum**, where fermented food is ground to a paste and much of its water content removed.

The **abomasum**, or true gastric stomach (similar to the human stomach) where pepsins break down proteins into amino acids.

The cow, like all herbivores, is not only vegetarian but also potential prey for carnivores. Her digestive system is designed to make use of grasses in particular, in bulk, and enables her to harvest, swallow whole and store substantial amounts during major grazing bouts and then slowly digest them when it is safe to do so by resting and regurgitating small lumps (cuds) of the food stored in the rumen, where enzymes have already begun their breaking-down process. You will be able to trace the way that a lump rises up her throat to be chewed comfortably so that it is mechanically broken down to help the digestive processes before she swallows it again and, after the briefest of pauses, brings up another lump for mastication.

A cow grazes with her *tongue*. She wraps it around the plants, draws them into her mouth and tears off the top growth. She has no top incisor teeth but breaks off the material using her lower incisors against a horny dental plate on the upper jaw. She has a full set of back teeth, top and bottom, which efficiently grind the plant fibres.

The first point to note, then, is that *cows eat bulky food*, essentially grassland plants, fresh or conserved, which are largely composed of cellulose and lignin — and it is these components which make grasses indigestible to humans. So the cow can utilize one of the most abundant natural ground-covering group of plants, which we ourselves cannot. People have always exploited the ruminant's ability to convert feedstuffs which humans cannot eat into products that they can use, be that food (meat, milk, blood), fibre (wool, hair), hide, dung (fertilizer, fuel), energy (muscle-power, gases) or any other by-products which can be made from bone, horn and offal.

Grazers and browsers are so valuable because they can give us all these things without taking the food from our own mouths.

The second point is that cattle need peace to *chew the cud* as an essential part of the digestive process. I have no idea what a cow thinks about while she does so but the original meaning of rumination is cud-chewing, not meditation.

The rumen is alive with bacteria; it is an internal symbiotic system, a veritable powerhouse of energy, and its efficiency depends upon the balance of life within: special bacteria break down cellulose in the roughage that forms the bulk of a cow's diet; other microbes release protein, starch energy and fatty acids. Levels of acidity of the rumen, depending on ammonia and salival sodium bicarbonate, are crucial to the well-being of the cow and the pH value needs to remain above 6.5 to avoid a range of physiological problems. To maintain appropriate levels, it is vital that the animal has ample fibrous food and not too much of the concentrates so heavily used to maximize production, especially in dairy cows.

With all the enzymic activity in the rumen, plenty of heat and gas are generated and the cow belches, quietly and fragrantly. Sometimes, however, something goes wrong: the cow cannot belch and the gases accumulate, blowing up the rumen almost to bursting point until it visibly balloons under the skin near the animal's left flank, tight as a drum — a condition aptly described as **bloat**, which can be fatal if not quickly relieved (see p.103).

NUTRITION

Cattle need food at two levels: for maintenance and production. **Maintenance** rations keep the body functioning properly — the repair and regeneration of body tissues, energy for breathing, blood circulation and normal daily activities. **Production** rations contribute to the growth of the foetus, the flow of milk, or the growth of a young animal.

It is because of human demands for increased production that so many things have gone wrong and that the subject of cattle-feeding has become complicated. For example, the overfeeding of protein, from whatever source, leads to health problems in cattle including loss of fertility in breeding animals, liver disorders, the excessive production of ammonia and high levels of urea in the milk. Moderation in all things, as ever. Never give cattle animal protein.

Without the emphasis on production, ration formulation is very simple. Maintenance requirements are usually met by feeding cheap bulk foods like grasses and root crops. Production at a low level can also be met by feeding maintenance foods if their quality is good. For unnaturally higher levels, however, an animal simply does not have the stomach (capacity or appetite) to eat enough bulk foods to provide the proteins and energy needed and is therefore fed cereals, which provide energy in a concentrated form.

From the cow's point of view, what counts is whether she likes the stuff and whether she has room for it — palatability and appetite. Cows *like* a handful of concentrates, regardless of productivity, and most respond readily to the rattle of concentrate pellets in a feedbowl — a useful handling aid even if pellet-feeding is avoided on principle.

FOOD

Let's forget all this formulation for productivity and concentrate on the well-being of the animal. On well balanced **grazing** during the summer, your cow will simply hoover up what she needs and should not want any extra feeding at all unless there is a drought, except perhaps for mineral supplements. Grass does not grow all the year round, of course, nor does it remain at a constant level of nutrition during the growing season. It is at its most nutritious when it is young, in full leaf and before it flowers. The **leaf** is more palatable and more nutritious; the **flower-stem** is more fibrous, more bulky and less digestible.

The grass plant grows rapidly in spring; it has another, lesser burst of growth in about September but its overall nutritional value has been dropping quite noticeably since midsummer. This point is discussed in more detail later, under the subject of grassland management and haymaking.

In winter, when grass growth is at a standstill and the remaining grass is little more than fill-belly, if that, other food is required. The main standby is in fact grass — the spring and summer flush that was preserved as either hay or silage when it was in surplus. As long as it was well made at the right stage of growth and properly stored, hay or silage should be all a low-production cow needs in theory during the winter and it can be offered perhaps twice a day. Now there are two problems, from her point of view: the stuff is all of a heap so that she does not have the luxury of choosing her 'grazing', and it's extremely boring to eat the same old stuff for months on end. If the quality is not good anyway, there is the possibility that the food will be deficient in some respect with nothing to counteract the deficiency. For these reasons, do the cow a favour and vary the menu. There is a wide range of suitable foods for winter cattle, and they divide into three broad categories.

- **Roughages** are high-fibre, bulky, dryish foods such as hay and feeding straw, which take some time to digest and are essential for rumen efficiency; they are cheap maintenance foods.

- **Succulents** are moist foods, much liked by most cattle but slightly laxative, such as roots, greens and silage; these, too, are cheap maintenance foods.

- **Concentrates** have a high dry-matter content, are rapidly digested; they offer a concentrated source of protein (legumes, oilseed) or of carbohydrates (cereals) and are relatively expensive production foods.

Each animal has its preferences but most of them welcome a chance to sample succulents in addition to their basic winter hay. Ideally these are grown deliberately as fodder crops and either eaten in the field, with access controlled by electric fencing to avoid wastage by trampling and dunging, or are 'cut and carted' to the animals in their yard or winter housing. The advantages of field feeding are that the cattle manure the land for the next crop and you are saved the labour of carting; but if the field is wet the crop will become soiled with mud, which can lead to scouring (diarrhoea), and if the weather is very cold the crop will be frozen, which can cause gut-ache.

Carted **greenstuff** can be offered whole, if necessary hung up to avoid being trampled. **Roots** are usually chopped before being fed. Try tempting with all sorts of sensible and tasty foods during the winter — go on, spoil them a little, but 'little' is the word: too much or too suddenly is bound to have indigestible consequences. Always introduce a new food gradually and never change a diet suddenly. But try out various goodies to see what your animals like and what suits them. Many cows appreciate sugarbeet pulp (dried, but soaked overnight to make it succulent and even more palatable) and some adore apples, for example. Mine always had a couple of apples chopped in half in their winter foodbin purely for the gourmet pleasure of them and I know of a heifer so addicted to apples that she will force her way through any fence, hedge or cowhouse door to get to an orchard and gorge. She once became decidedly tipsy on rotting apples . . .

Cereals are important sources of carbohydrate energy. Most people give their cattle rolled barley but in fact oats are preferable to barley or wheat: they have a higher proportion of digestible fibre and fat and a lower proportion of starch. They cause fewer digestive upsets, and they help to produce high-butterfat milk with a lower proportion of saturated fat. Cereal straw is sometimes fed as a cheap substitute to eke out the hay ration: use spring barley, which is softer to feed than winter barley or winter wheat straw (the latter is often uncomfortably hard and indigestible), or if you can achieve it, use spring oats straw harvested by binder at an immature stage, when it is probably the best of all feeding straws. Straw is more suitable for store cattle, fattening cattle and suckler cows than for particularly productive milking cows but it does provide the digestible long fibre necessary for rumination and for the production of good butterfat levels in the milk, like any roughage. But they'd prefer hay!

THE BASIC MENU

GRASS/CEREAL FORAGE	ROOTS	LEAFY FORAGE*
Grass	Mangolds	Kale
Hay	Turnips	Forage rape
Silage	Swedes	Fodder radish
Clover	Fodder beet	Stubble turnips
Barley straw	Potatoes	Cattle cabbage
Oat straw	Carrots	Sugarbeet tops
Forage maize		Fodder herbs
Triticale		
Silage mix (oats, vetches, field beans, forage peas)		SEAWEED MEAL is useful for calves

BEWARE of choking on roots.

BEWARE of feeding: fresh mangolds, green potatoes, unwilted beet tops

BEWARE of feeding in excess: kale, rape, fodder beet, sugarbeet pulp

* **Forage** is bulky food (as opposed to concentrates) in which the energy component is relatively diluted, as in hay and silage.

How much food should you provide? Well, some animals are sensible eaters and, given the chance, will select what they need for a balanced diet and will stop eating when they have eaten enough, but others are just plain greedy. Be guided by the individual animal's condition as much as by its appetite: cows in particular should be fit, not fat, especially in preparation for calving. Some cows seem to be able to eat hugely and put it all 'into the bag' in the form of more milk but others put it all 'on the back' and get fat without adding a drop to their milk yield. This is usually a genetic fact of life, by the way, not just a perversity of character.

Treat each animal as an individual. On average, if you are not trying to force the production of a lot of milk, a **Jersey** cow would need about 6-7kg (13-15lb) of good quality hay a day for maintenance only, or up to 10kg (20lb) to produce about 5 litres (9 pints) of milk a day. A **Friesian**, being bigger, would need 8-9kg (17-19lb) of hay for maintenance and about 0.8kg (2lb) of extra hay for every litre of milk she produces.

It's all very well talking in kilos, but who has a scale large enough? A **bale** of hay might weigh anything from 15 to 30kg (30-65lb) but you'll soon know how many 'sandwiches' to give your cow. Fluff them out in racks or haynets to keep the hay off the ground, unless your outdoor cattle are hungry enough to eat it rather than trample on it.

To replace part of the winter hay ration with other foods, use the hay equivalent table as a rough guide on how much of an alternative food to give a low-yielding animal, but make sure you give enough long-fibre food to help good rumination.

Hay equivalents

1kg (2lb) of good hay could be replaced by:

	3kg (6½lb) of good silage
or	2kg (4lb) of good feeding straw
or	5kg (11lb) of swedes or mangolds
or	4kg (9lb) of kale or cabbage
or	0.5kg (1lb) cereals

Take care in feeding alternative fodder crops. Roots such as fodder beet, sugar beet, mangels, swedes and turnips are full of energy but do not offer much protein and should be limited to a maximum of about 3kg (6½lb) *dry matter* a day (which, as they contain about 80-90 per cent of water, still means quite a lot of wurzels!). The green tops of beet, swede and turnip are higher in protein than their roots but again, limit feeding to 3kg (6½lb) dry matter a day, and *wilt* sugarbeet tops for several hours before feeding. Kale and cabbage can affect an animal's fertility if fed in excess: kale in particular should be limited.

Minerals and vitamins

If the food you offer is deficient in certain minerals, or conversely overloaded with them, your animals might suffer certain metabolic problems. The best way of counter-balancing such deficiencies is to offer a good variety of foods and, if necessary, import produce from other areas with a different balance of soil minerals.

If you grow organic crops properly you have a better chance of producing a good balance of minerals in the rations. In addition, **herbs** are a good source of minerals and play an important part in the life of a grazing animal. They should be included in the grassland (as described later) or available for browsing from the hedge and headlands. You can also grow certain important herbs as fodder crops, especially borage, comfrey and chicory. These useful plants can easily be grown in the garden and fed after being cut.

Vitamins are less of a problem as most of them are produced by the rumen's bacteria anyway except for the fat-soluble vitamins A, D and E. The latter should all be available in adequate proportions in a well-balanced ration but vitamin D is the 'sunshine' vitamin: it has a close association with sunlight and there is nothing like a touch of winter sunbathing to cheer up a jaded cow! Vitamin sources include:

- Carotene for vitamin A (found in carrots, green vegetables, milk, cod-liver oil).
- Yeast and cereals for vitamin B.
- Milk and cod-liver oil for vitamin D.

Organic standards in feeding

One of the main principles in organic livestock farming is that all — or least as much as possible — of the animals' food should be produced on the holding, partly so that you know exactly how it has been grown and partly to retain the energy cycle within the holding. It will also make you take an even more responsible attitude to your own land, and make you appreciate the value of hard work over money!

You should therefore aim to produce all the forage your animals need, which means very careful management of the grazing to ensure that you can make adequate winter stocks of hay or silage for a start; and well planned timing of fodder crops like greenstuffs and roots. Be self-sufficient as far as is humanly possible.

It might be more difficult to grow your own grain as the basis of concentrates for production, if they are necessary, but at least try to buy the grain from a known, reliable organic source (the organizations listed at the back of the book can help), though it will be relatively expensive as there is a premium on organic grain which could be sold for human consumption. Avoid using those ever-so-convenient bagged concentrates or manufacturers' mixes which contain you know not what; the lessons of the BSE epidemic might or might not have been learned by the feed producers.

Organizations which try to set standards for organic farming allow the limited use of feeds (especially concentrates) from non-organic (conventional) sources but preferably as a last resort, partly because so many of the national herd's physiological problems arise from the misuse or overuse of them. In general the allowance is up to a maximum of 10 per cent of the whole ration on a dry-matter basis. The Soil Association adds that at least half the ration must be from organic sources and that the non-forage proportion (including, for example, sugarbeet pulp as well as concentrates) must form no more than 40 per cent of the daily dry-matter intake. However, the better the quality of home-grown forage, the less likely it is that you need to feed concentrates at all.

Additives such as growth promoters and other hormones (for example, BST — bovine somatatrophine used for boosting milk yields), antibiotics, medicated feeds, urea and so on are prohibited in organic systems, as are all feed supplements of animal origin for any ruminant. Naturopathic **probiotics** need specific approval from the Soil Association but should not be necessary (even if you are convinced of their effectiveness) if your basic husbandry is sound — let the rumen develop its own microflora. (Probiotics are naturally occurring bacteria, but artificially introduced. They are said to help promote growth and food conversion, especially in young animals, by playing a digestive role or actively controlling pathogenic bacteria, crowding them out.)

GRASS

Grazing provides the bulk of a cow's diet. Grasses and herbs make the cheapest meal on the menu and are also her staple preference. Grassland gives the best and most economical results if it is treated like a crop rather than a shaggy tablecloth in need of a trim. The field should be *managed*, even at the level of house-cows and store cattle, for the sake of the animals and for the sake of the land.

- **Grazing** is the act of feeding from a growing crop, usually, but not necessarily, grass.

- **Grassland** is essentially an area, used principally for livestock feeding, in which the main species are grasses.

- **Pasture** is a field containing a mixture of grasses and herbs for livestock and is usually permanent, or long-term grassland which has not been ploughed for reseeding for several years.

- **Leys** are grasslands deliberately sown with selected species as a crop for grazing or winter-forage conservation, usually on a temporary basis for a set number of years from one upwards.

- **Rough grazing** is low-value unimproved grassland, usually on uplands (moors and hill farms), on which grasses are not necessarily dominant and might be quite overwhelmed by, say, heather, in which case it is more suitable for sheep than cows, though beef cattle can find a living if the heather leaves room for grazing plants.

Permanent pasture is the most practical grassland for a smallholder: it is labour-saving, it can build up a good range of plants and a firm well knit sward, it provides a good habitat for wildlife, and usually, it's already *there*. It takes a summer drought to bring home the importance of well established pasture stocked with a good range of plants. Their root systems, built up over the years, are better able to withstand drought than temporary monocrop grasses, and clover in particular defiantly stays green when all around it the grasses are withered and brown. But pasture does need some management, partly to maintain its species variety, partly to utilize the various stages and degree of growth, partly to keep the land and the animals healthy.

The growth rate and goodness of grass

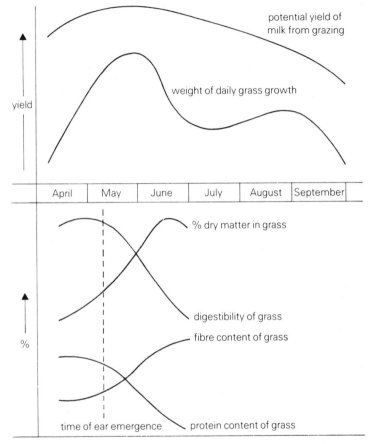

My first attempt at sowing a new pasture for house-cows was in September 1975, which was promptly followed by the long summer drought in 1976, when only the clover and the chicory in the mixture stayed green. I was reduced to grazing two cows on the blessedly damp greenway by the cottage and on every scrap of headland around a neighbouring farmer's crops, and I was deeply thankful that my dear, naive creatures were firmly convinced that my baler-twine 'fence' was electrified — they never once tried to nibble the tempting greenness of the corn crops.

The **ideal** pasture contains a wide variety of edible plants including grasses, legumes and herbs, so that:

- The cow has a choice of tastes and textures
- There is a better chance of a balanced diet with adequate minerals
- There is always something to eat, with different plants at their best at different stages
- Soil fertility is maintained by contributions from legumes and herbs
- The cow can selectively graze medicinal herbs
- The plant diseases and disasters of monoculture are avoided.

PASTURE PLANTS

MAIN PASTURE GRASSES:

Cocksfoot	Italian ryegrass	Rough-stalked meadow grass
Timothy	Perennial ryegrass	Smooth-stalked meadow grass
Meadow fescue		

LEGUMES:

White clover (essential, long period of good digestibility)
Red clover (productive but short-lived — and beware of bloat)
Alsike, tritifolium, trefoil, vetches
Lucerne (for hay/silage but not for fresh grazing)

CHICORY: deep-rooting, drought-resistant, full of minerals, useful in winter too, a good tonic.

SALAD BURNET: deep-rooting for calcareous soils, well liked, medicinal properties.

RIBGRASS (PLANTAIN): very palatable, high in minerals.

YARROW: extensive root system, drought-resistant, medicinal, but can become too dominant if not close-grazed.

SHEEP'S PARSLEY: high iron and vitamin content for sheep but tends to be biennial and does not persist.

AND ALSO (but some taint the milk):

Cat's ear	Caraway	Borage	Comfrey	Tansy
Rosemary	Marjoram	Sage	Dill	Fennel

To maintain species variety:
- Cut your hay later rather than earlier, letting different species flower and set their seed.
- Graze limited areas intensively and then give each patch a good long rest to allow slow-growing species to recover and flourish.

Different species grow at different rates and at different times, and the art of pasture management is to make use of those differences to ensure that good grazing is always available and that the inevitable seasonal surpluses are not wasted but are conserved for feeding in winter as hay or silage. As explained

MILKER'S HERBS

To increase milk yield

Anise	Marshmallow
Balm	Blue Melilot
Borage	Milkwort
Clover	Sage
Comfrey	Sow thistle
Fennel	Speedwell

To increase butterfat

Carrot	Maize
Chervil	Marigold
Elderflower	Oats
Linseed	

Useful milk curdlers

Butterwort	Nettles
Blue Melilot	Sorrel
Lady's bedstraw	

Butter/cheese problem-makers

Mint (won't clot or churn)
Sorrel (won't churn)
Wood sorrel (won't churn)

Milk tainters (note that some are yield/butterfat increasers!)

Anise	Garlic/Onion	Ox-eye daisy	Water parsnip
Buttercup	Hedge mustard	Pennycress	Wild radish
Camomile	Ivy	Sage	Wood sorrel
Chervil	Knotgrass	Sorrel	Wormwood
Cow wheat	Knotweed	Tansy	Yarrow
Fennel	Marsh marigold	Turnips	
Fool's parsley	Mint	Watercress	

earlier, grass grows fast in spring, when most species of grass are at their most nutritious, but some grass species are early shooters, while others are late starters that continue to be palatable for longer.

As well as grasses, every pasture needs legumes — especially **clover**, to fix nitrogen in the soil and avoid the need for artificial fertilizers. There are drawbacks to clover, especially in view of its ability to take over when grass is curtailed by drought. Clovers and other legumes are sappy plants, which means they need more drying in making hay or silage; they can also, in excess, cause bloat; and certain varieties of red clover contain high levels of oestrogen, which can affect an animal's fertility.

Bloat warning: Over-indulgence on new spring grass by animals used to winter rations, or on clover (especially in its late-season flush), could cause bloat in cattle. This is very dangerous — see Health chapter for symptoms, prevention and treatment.

Herbs of various kinds are not simply to tickle a cow's palate; they are in fact essential. As well as being rich in minerals, many have medicinal or tonic properties to keep the cow in good health. Some are also deep-rooting and therefore drought resistant, which can be a great blessing in dry summers, while others bring welcome greenstuff in winter as well.

In the context of pasture, most **weeds** are not weeds at all but useful ingredients contributing to the whole. However, some do need controlling, either by mechanical removal or by recognizing that their presence indicates certain factors which can be avoided or controlled. **Thistles**, for example, tend to appear in fertile but undergrazed pasture and can gradually be eliminated by persistent cutting; top them in June before they set seed and again in August, and keep doing so until they give up (regular cutting for hay and silage helps, of course). **Docks** also like fertile ground, especially well dressed with slurry or nitrogen, and favour areas which have been poached or compacted: avoid soil compaction to deter them in the first place, or top them to prevent seeding and keep livestock in the field to control them by constant grazing (though many animals avoid dock leaves). **Rushes** are a sure indication of bad land drainage, and **buttercups** also prefer damp, acid soils, especially where the pasture is undergrazed or the field has been poached.

Ragwort is poisonous to livestock. Animals usually avoid it during grazing if it is standing and growing but it becomes even more poisonous when it is cut and wilted — and less recognizable. It remains toxic, or even lethal, in hay. You must remove ragwort by pulling it, by hand, and making absolutely sure you take out every tiny scrap of root to prevent regrowth.

Pasture management

The management of permanent pasture involves topping, harrowing and controlled grazing.

- **Top** the field if the grazing runs away from the animals: top in June to about 8-10cm (3-4 inches), and again later if necessary. Topping keeps the grazing palatable by encouraging fresh growth and removing rank, untouched herbage; it also prevents unwanted weeds from setting seed.
- **Chain-harrow** the field in the spring to scatter cowpats and molehills, to drag out dead and matted vegetation, and to aerate the surface of the soil.
- **Roll** the field in spring to settle frost-lifted roots and to even the surface if you intend to cut for hay or silage.

Manuring is best done by the grazing animals; their dung and urine (combined with nitrogen fixed by the clovers) should be adequate compensation for the soil nutrients removed during grazing.

Now for the **grazing**. The main features of good grazing management are:

- Maintaining a balanced sward
- Providing good food all year round
- Stocking at rates appropriate to the grass and to the manuring of the land
- Encouraging good aftermath growth (**Aftermath** is regrowth after an area has been cut for hay or silage and offers fresh grazing on a patch which has not been grazed earlier in the season.)
- Controlling internal parasites

Stocking rates depend on judgement and system. The aim is to make best use of available grazing, while allowing adequate areas to be set aside for winter-fodder conservation. The choice lies between continuous and set stocking or rotational grazing.

- **Continuous stocking**: Cattle have access to the whole area for the whole season, but the number of cattle on that area can vary according to the state of the grazing.
- **Set stocking**: A fixed but low number of cattle graze a fixed area for the whole season.
- **Rotational grazing**: Cattle are given access to a limited area to eat it down rapidly, then move to a fresh area and let the original recover for several weeks before it is grazed again.

Continuous and set stocking make life much easier for you — as long as the cattle have plenty of space they simply get on with it, choosing where they want to graze and when. However, if the area is too large the grass will grow away from them and become less valuable and less attractive. Worse, if the area is too restricted you run the risk of a bad build-up of internal parasites.

One of the aims of pasture management is to keep the height of the sward at optimum levels both for the cattle and for the plants, neither overgrazing nor undergrazing it. Plant leaves need adequate light and are soon shaded out by the developing flower stems and heads if the sward is not either grazed or mechanically topped. The ideal sward height for cattle is about 8-10cm (say 3-4in).

Rotational grazing avoids some of the problems. It helps to control parasites, it gives fresh grass a chance to grow and, by letting the pasture rest, it allows time for insects to disperse the cowpats. The main choices for rotational grazing, even on a very small scale, are: paddock/strip grazing or stock rotation.

For **paddock grazing**, the whole grazing area is divided into smaller paddocks, often by means of temporary electric fencing. For **strip grazing**, an electric fence is run across the field to let animals graze up to it, then the fence is moved forward little by little to give fresh grazing, perhaps with a back fence to keep them off the grazed section and let it recover. (This system is also used for grazing fodder crops like kale in winter.)

In **stock rotation**, different *species* of livestock are grazed in succession over each paddock or field. Different species have different methods of grazing: sheep nibble the sward down close, in contrast to cattle who need longer stuff to wrap their tongues around. Put cattle on the pasture first, and then let sheep tidy it up after them, always in that order. Then let the grass rest for a while.

Parasites

Parasites can build up to become a major burden where grazing space is limited and the best remedy is always to have more than enough ranging for the number of animals you keep, so that each area can be rested at intervals for several weeks at a time or, even better, cultivated for a crop for a season. Overstocking is asking for trouble and it is far more sensible and effective to control parasites by proper management of the grazing than to have to treat livestock with worming remedies.

Controlled grazing (paddock/strip or stock rotation, as just described) helps as a secondary measure if you do run short of space — the techniques help to break the parasites' life-cycles as they are often host-specific and cannot complete their cycles if the appropriate host is absent at the appropriate time. For example, many sheep parasites do not pose a threat to cattle, and vice versa.

Most creatures play host to various internal parasites to some extent and a healthy, mature animal can tolerate them unless the burden becomes excessive or unless the animal is already a little weak. **Young animals**, however, have little inbuilt resistance to parasites and *must* be given 'clean' grazing until they gradually acquire a degree of tolerance, which can take several months, though calves running with their mothers are better able to cope with parasites than calves on their own. Ideally, only let young animals graze on pasture which has never been used by adults, or at least not during the current growing season, i.e. make sure that there has been *at least* a winter's break in between.

Some parasites also require intermediate hosts (liver flukes, for example, spend part of their lives parasitizing snails) and can be deterred if the actual environment is altered to make it unattractive to the intermediates, perhaps by land-drainage. The same applies to external parasites such as ticks, whose attentions can lead to several feverish diseases and which also, indirectly, cause redwater in cattle (symptoms include red urine). Ticks can be discouraged by ensuring that rough grazing does not grow away and that bracken is eradicated. On the whole, however, external parasites are much less of a problem for cattle than internal ones.

The symptoms and treatment of disorders caused by internal parasites are described in the Health chapter.

Hay and silage

Hay and silage are conserved grass: they make full use of the surplus grass your cow has been unable to keep up with early in the season by storing its nutrition and energy for her use in winter. All that harvested sunshine . . .

- **Hay** is conserved by drying in the field: it has a high dry-matter content.
- **Silage** is conserved by pickling; it is fermented in airtight acidic conditions and remains succulent.

Many farms no longer make hay but rely heavily on silage instead, and there are factors for and against both. Cows will eat either. Organic farmers often compromise, making a limited amount of hay for their youngstock perhaps but converting most of the surplus grass into silage. For smallholders, hay is probably the better option. You would need about a tonne per cow to see you through the winter.

Hay smells delightful, is easy to handle and feed and store, is well liked (if well made) by all livestock, does not pollute watercourses and gives you an endless supply of baler twine! But you will possibly take only one hay crop, two at best. However, it is cut later than silage, which gives ground-nesting birds a better chance and allows a wider range of plants to seed themselves and maintain variety in the sward. You need good weather for haymaking — at least three consecutive dry days and preferably sunshine. It is perfectly practicable to make hay by hand or with the help of a horse (if you can find old machinery); if you need the help of a contractor, however, you are bound to lose the weather-window because everybody wants contractors at the same time.

Silage can be first cut very early, when the grass is at its most nutritious, and it might be possible to take several more cuts during the season. You do not need a spell of fine weather; the crop is only wilted in the field rather than dried in the sun and can even be cut wet, though not when it is actually raining. It can be removed from the field within hours of cutting. Silage needs very careful storage in anaerobic (airless) and leak-proof conditions; the run-off of silage liquid can be a serious water pollutant. Silage is a succulent fodder, which cattle welcome as a change from the dryness of hay, but its smell can be far-reaching and some people don't like it. It is a pickled food, and the pickling often goes wrong. It is also bulky to handle and most people take the animals to the silage rather than vice versa unless they have adequate machinery.

Whether you decide to make hay or silage, you must shut up part of your field for about three months whilst the grass grows. You need to remember that by removing the grass crop from the field you are also removing goodness and it is important to return nutrients to the soil by some means. If grass is grazed in situ, of course, the land is dunged by the animals; if the crop is removed and fed to housed or yarded livestock, it is not. Compensate for the loss by rotating conservation areas. If a patch is set aside for haymaking or silage one year, make sure it is fully grazed next year to restore fertility, and make a practice of grazing aftermaths.

Farmers who rely on silage usually cut it from temporary leys deliberately planted with only one or two grass species which grow at their maximum rate at a season convenient for the silage cut — all the plants in the field are ready for cutting at the same time, either for a very early crop in May or at intervals during the summer. If you are taking silage from mixed pasture and cutting it several times, you run the risk of limiting the recovery of certain valuable species and you will certainly deprive them of the chance to seed themselves. It might be necessary to make a deliberate point of planting herbs on the headlands so that they remain uncut and can grow unmolested until cattle graze them.

Whether it is to be hay or silage, the quality depends on:

- The quality of the grass at the time of harvest.
- The efficiency of the conservation process.
- The efficiency of handling and storage.

As soon as grass has been cut, it begins to deteriorate and die. While it is dying it is using up reserves of energy; the trick is to 'kill' it quickly before those reserves are lost and the food value diminished. The process is halted by desiccation or by lack of air.

To make hay
The most nutritious hay is made early in the season, just before the flowerheads open, when its protein and starch are at relatively high levels but the fibre content is relatively lower than

in hay cut in full flower. The most abundant hay is the later cut, but its quality is lower so that you need to feed more of it. Take your choice!

Cut only as much as you can successfully make before the weather breaks. Drop it in windrows and let it dry as quickly as possible in the sun: fluff it up occasionally to let the air circulate, expose hidden layers and promote even drying throughout, but be aware that it becomes increasingly fragile as it dries — don't be violent with the fluffing or you will shatter the valuable leafmatter and leave it in the field — until dry enough to gather up and store. If rain threatens before you are ready to store, heap the hay into cocks or get it all under a large pole-barn to finish drying before it is stacked or baled.

To make silage
The best silage is cut early, before ear emergence — it can usually be cut about a month before hay would be ready. You want leaf, not stem. Wilt the cut crop in the field for 24 hours if possible (bring it in rather than leave it out in the rain). Then transfer it to a heap or plastic bags — a situation in which all air can be excluded for the sake of the pickling process. Compress the crop within this containment, squeezing all the air out of it and then covering with plastic sheeting and sealing it thoroughly so that it remains airtight. Then just leave it to cook until the winter.

4. The Next Generation

The wild cow has certain behaviour patterns that her domesticated cousin would no doubt gladly follow if she had the opportunity, in spite of several thousand years of control and genetic manipulation to modify her character. To understand the mother cow's instincts and give her scope to express them, start by considering the wild or feral animal.

THE COW'S CYCLE

The wild herd of cows and their young are separate from the bulls but the groups remain within range of each other. A cow who is suckling a calf tends not to come on heat (become receptive) again until the calf is less dependent upon her, which is probably by about three months old. Then there are three main indications to the distant bull that, somewhere, a cow is ready for mating (the common term is that she is **bulling**): he will detect her **smell**, hear her **calling**, and he will see, from afar, that cows are **mounting** each other.

Smell is an important part of mating behaviour and the smell of a cow on heat triggers much mutual sniffing and chins-resting-on-rumps among the cows within their herd. That means quite a lot of general disruption, which alerts the bulls to the situation. The bulling cow wanders about among the cows, disturbing everybody and being a nuisance as well as a source of interest and she draws attention to herself across the landscape by attempting to mount other cows and by standing while other cows mount her. A cow who is not on heat will not tolerate being mounted by anybody, cow or bull, and moves away from such attempts but is well prepared to mount a cow who *is* on heat, and the latter stands still while she does so.

This behaviour pattern, originally to attract the attention of distant bulls to the fact that somebody in the cow herd is in the mood, persists in domestic cows though an isolated house-cow might have difficulty in expressing it. Be warned that it is not unknown for a bulling cow to attempt to ride a heifer young enough to be flattened or even a familiar human, equally flattenable, and that in precocious breeds like the Jersey a heifer might start bulling when she is only a few months old and far too young to be mated.

Domestic cows do not seem to be seasonal breeders and can conceive at any time of year, though their heat periods tend to be shorter and less obvious in winter. There are several indications that a cow is ready to mate, but not all cows show all the symptoms, and indeed some owners swear that their cows show no signs at all. However, if you know your cow's normal behaviour well enough, you will detect subtle changes in her mood and body language anyway, even if there is no obvious physical symptom of heat. Things to look out for in a bulling cow are:

- **Calling** persistently for a day or two for no apparent reason, seemingly to a distant listener, with a slight change in tone from her usual voice to a minor key.
- **Mounting** other cattle and (the important distinction) allowing them to mount her.
- **Reacting** archly if you press your hands firmly on her back above the root of the tail.
- **Raising** her tailhead slightly.
- Showing a '**bulling string**' — a clear mucous discharge from a slightly swollen vulva, or a similar discharge slightly streaked with blood a couple of days after bulling.
- Being generally more **restless** than usual, a little more

aggressive to other cattle with extra head-to-head pushing, more mutual grooming and perhaps more affection towards a trusted human.

- Showing a brief drop in her **milk yield,** perhaps by about 10 per cent for a day or two, because she is too busy bulling to eat properly.

The bulling period is short. On average the active signs are observable for perhaps two days and are repeated three weeks later if the cow has not become pregnant in the meantime. Again, individuals vary; some are obviously bulling for the full two days, some show only the briefest indication for an hour or two; some bull with absolute regularity every 21 days, others vary but normally the cycle is within 18 to 23 days, though a few cows might have as little as 8 days in between. The first bulling period after calving might occur from within a few days to several weeks but most people do not try to get a cow pregnant again until about three months after calving, so that she has a calf a year (the average pregnancy lasts nine months). Never be tempted to put a cow in calf again sooner than two months after calving. Nor is it necessary to calve a cow every twelve months except in a commercial situation; take her off the production line and give her a rest, though some cows seem to have a mind of their own on this and manage to calve annually on almost the same date each year, given the option of choosing when to accept the bull.

An experienced bull knows exactly when a cow is ready to be served and anyway mounts her several times so that he is bound to happen upon the peak of her fertility if he is allowed to run with her. With suckler cows and groups of mature heifers, the usual practice is to let them run freely with the bull so that they can all get on with the mating business in their own time. With dairy cows, or any other cow who, for whatever management reason, is not running with the bull, it is up to the humans to perceive that she is bulling and to take her to the bull for a few hours.

However, it is only economical to keep a bull if you have enough cows to merit his expenses (including special, sturdy housing) or in very special cases concerning rare breeds. Most dairy farms, and most house-cow owners, find it much easier to make a quick 'phone call to the local AI centre and have the cow artificially inseminated. It might deprive her of the pleasure of a

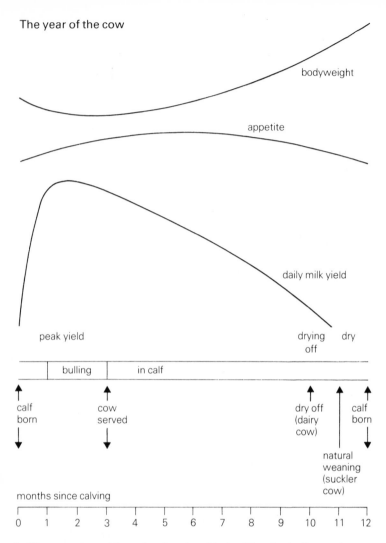

The year of the cow

bodyweight

appetite

daily milk yield

peak yield

drying off   dry

| bulling | in calf |

calf born

cow served

dry off (dairy cow)

calf born

natural weaning (suckler cow)

months since calving

0 1 2 3 4 5 6 7 8 9 10 11 12

bull's attentions (though why should she like the bull you force upon her anyway?) but it does give you a much greater choice of sires for the calf and only costs a few pounds at the time of insemination rather than tying up hundreds or thousands of pounds in your own live bull.

A major drawback of AI is that the timing must be right. You must be quite certain that the cow is in fact bulling and has reached the '**standing heat**' stage (prepared to be mounted without moving away). If you first notice that a cow is bulling in the early morning, have her inseminated that evening; if you first notice her in the evening, have her inseminated the next morning. If you 'phone the AI centre before about nine in the morning, a trained operative will arrive that afternoon and the cow should be tied up waiting for what is a quick and painless procedure, if a little undignified. Have her confined in such a way that her movements are restricted by side rails but so that the operative has easy access to her back end, and stand by to hold up her tail.

If the timing was not quite right, or she simply doesn't 'take', the cow will no doubt come bulling again three weeks later. Some people have their cows inseminated twice during the same heat to give a better chance of their holding to service (i.e. conceiving a calf).

Most people use the AI services of their local Milk Marketing Board. Others use private companies for special breeding and some companies will hold stocks of frozen semen from a farm's own bull (so that it is still available after his death) while others deal with certain rare breeds and help to ensure that there is less inbreeding where breed numbers are low.

Oestrus, service and ovulation

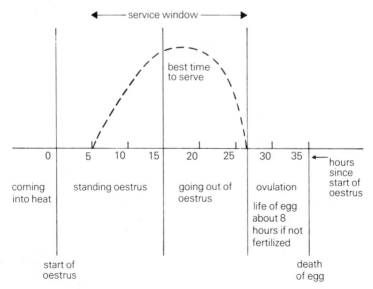

Choice of bull

At the local centre you can simply accept the bull of the day or choose more carefully as part of a deliberate herd improvement plan. You can choose from quite a wide range of different breeds, too: you might want a pure-bred calf, especially if it's a heifer to raise up as a milking cow — but there is no guarantee your cow will always oblige by giving you heifer calves. What *will* you do with a purebred Jersey bull calf (pretty useless for sale as beef)? People often prefer crossbreds for hybrid vigour or to put more meat on a dairy cow's offspring. But do take care not to choose a type of bull whose calf will be big enough or bulky enough to cause calving difficulties. A Hereford bull is a good sire on almost any breed of cow if you want beefier calves and he *always* passes on his characteristic white face to all his progeny — a trademark much appreciated by buyers wanting a good beef calf. Another excellent beef sire is the Aberdeen-Angus, who passes on his top-quality beef characteristics and is small enough, with a tapered polled head, to be an ideal partner for a heifer's first pregnancy or for a small breed of cow who might have calving problems put to a bigger breed of bull. The Murray Grey, an Australian breed, originated from Aberdeen-Angus and Shorthorn and is another possibility. Many other beef breeds are described in the breed tables in Chapter 2, and it should be noted that it is traditional to describe a crossbred showing the bull's breed first. For example, the useful Longhorn × Welsh Black suckler cow has a Longhorn father and a Welsh Black mother, while the Luing originated as a Beef Shorthorn × Highland.

It is increasingly common for commercial farmers to use Continental bulls, which tend to produce a great deal of very lean meat but many of them sire calves which are so big and muscular that the poor old cow inevitably has problems giving birth. The most common Continental beef sires currently used in the UK include, from France, the Charolais, Limousin, Blonde d'Aquitaine and Salers, the hideously double-muscled Belgian Blue, the sturdy dual-purpose Simmental of the Alps, the yellow Gelbvieh from Germany, the handsome Alpine Pinzgauer and the newly fashionable big white or grey Italians like the Romagnola and the very tall Chianina.

A few people become involved in embryo transplants but this is highly specialized and expensive breeding for commercial concerns or for those who have a mission to conserve rare breeds. It can be very time-consuming, very frustrating and, frankly, not very pleasant for the cow, whether she is the one who donates her valuable eggs or the common-or-garden cow who is implanted with a superior cow's embryo.

PREGNANCY AND CALVING

The average pregnancy lasts about 40 weeks (280 days) but, again, individuals vary and the breed of bull also affects the length of pregnancy to some extent, though it will only be a few days either way. Keep a record of when the cow was inseminated, if you know the date, so that you are prepared for the calving in due course. However, quite a few potential pregnancies fail for one reason or another, especially in the first three weeks and you should avoid putting the cow under undue stress during that crucial period: don't be unkind to her, don't disrupt her group, don't change her routine and don't change her diet. Otherwise, as long as she is basically fit, properly exercised and sensibly fed, a cow does not need any special attention during her long pregnancy — it is a natural state of affairs after all, not an illness — but do make sure that she does not become too fat or she will probably have calving problems. The motto is, 'Fit, not fat.' If she doesn't come bulling again after three weeks, she's probably pregnant, but you can get your vet to come and test for pregnancy if you want confirmation.

Calving

The wild cow withdraws from her herd shortly before calving, so that she can give birth privately and in peace. She will clean up her new calf, devouring the afterbirth to fool predators, and suckle the little one before leaving it 'lying up' like a fawn, hidden in long grass or a thicket, while she goes off to graze with the herd. She returns to suckle her calf perhaps every 5 hours in the first couple of days and perhaps every 3 hours over the next few days, each suckling bout lasting for about 10 minutes at a time, and she leaves it to sleep in between each meal while she goes back to the herd. The new calf is *used to being alone* except when it is being suckled and groomed by its mother.

After a few days the calf is ready to join the mother's herd and then it tends to spend most of its time with other calves. They soon establish a routine with all the calves lying down in a 'crèche' group to rest for a few hours after suckling, while their mothers graze at some distance until a calf gets hungry and starts calling for mum. In typical cow-herd fashion, once one of them is suckling the others follow suit and soon all the cows are attending to their calves' demands. Although the calves begin to nibble vaguely at the grass almost as soon as they have joined the group, they continue to demand their milk for several months and are still suckling three or four times a day at six months old.

A domestic cow also has that urge for privacy when she is ready to calve and, given the chance, will take herself off to a quiet place some distance from other cattle and people. On the whole she is much better left to get on with it in a place of her choosing, where she feels safe and unpressured. Unless the weather is bad (which, from the point of view of the cow and her calf, means chilly, wind-driven rain) it is usually much healthier for the calf to be born out in the field, provided that there is somewhere to shelter and no obvious hazards (including loose dogs). The first you will probably know of the whole business is when you notice that the cow suddenly looks slimmer and rather pleased with herself. You might not see her calf unless she's suckling it!

Some cows calve as readily as rabbits drop their kittens but there is always a chance of a malpresentation or some other calving problem and you should at least be aware of the imminence of calving. Be prepared to keep a *discreet* watch on her — but too obvious observation could put her off the whole idea. In fact most cows calve at night, often in the small hours of the morning. In the case of a first-calf heifer you really must be around in case her inexperience lets her down.

If you decide to calve indoors, either because of the weather or because you anticipate calving problems (neither you nor the vet will enjoy trying to find a cow in difficulties out in a dark, wet field), give her plenty of time to get used to the idea. Make a habit of bringing her in daily for several weeks before the due date so that she will feel at home inside — especially if she is a heifer unused to being handled. The calving quarters must be very clean indeed, well disinfected and well littered with fresh, bright straw bedding, with ample space for the cow to move around in during her labour. It is important that she is familiar with her surroundings for several days before calving.

The **signs of calving** within days include:

- A general change of mood — perhaps a certain dreaminess, gentleness and affection, or an air of restlessness and more mooing than usual.
- Withdrawal from social conflicts, and perhaps physical withdrawal from the whole herd.
- Udder 'bagging up' perhaps 3-4 days before calving (individuals vary); it begins to swell with milk. Watch out for *undue* discomfort but only relieve pressure *in extremis* — it is best left alone. Consult the vet if you're worried.
- Swollen, flabby, pink vulva two or three days before calving, sometimes with a slight, clear mucous discharge.
- Pin-bones 'dropped': pelvic ligaments on either side of tailhead relax to let the pelvic outlet stretch during the birth. You can slip the edge of your hand into definite cavities between the tailhead and the adjacent bones. Calving is probably a day or two away.

The **immediate signs** of birth include an increasing restlessness and fidgeting, an enhanced awareness of the surroundings (eyes, ears and nostrils on the alert for predators) and occasional twinges of slight discomfort, rather like indigestion, with a little brief straining now and then.

The birth
Initially the cow will probably lie down as her straining becomes more powerful. You might see signs of the breaking of the first water-bag, discharging urine-like liquid. She might have a little rest for an hour or so and then continue with spasmodic straining until the first tip of the calf begins to protrude — literally the tips of its two front feet, soles down. Next comes its tongue, then its muzzle and head, then its shoulders and chest. The cow will probably stand up so that the rest of the calf quickly slips out, the umbilical cord breaking as the baby slides to the ground.

That is how it *should* happen and, except in the case of malpresentations, it is usually best to let the cow get on with the job herself, in privacy, even if it takes time. Don't interfere unnecessarily. *Experienced help is necessary* if:

- A properly presented calf (front feet first) has had its tongue and nose showing for an hour or so with no further progress.

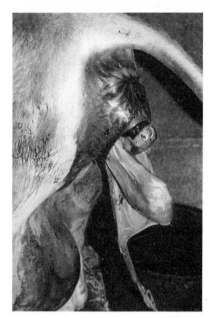

The normal calving sequence:
First sign of the calf, forefeet showing

. . . followed by muzzle.

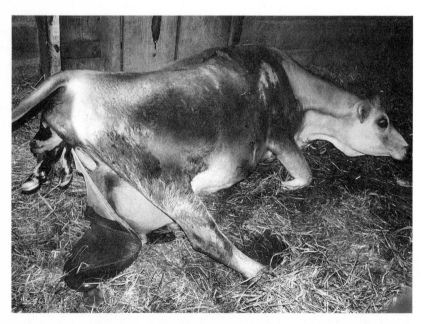

Delivery continuing apace.

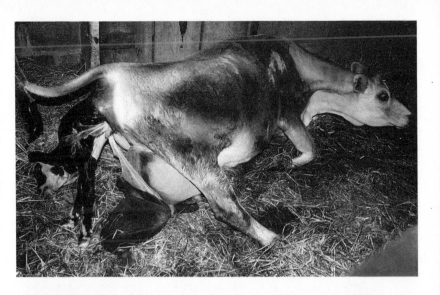

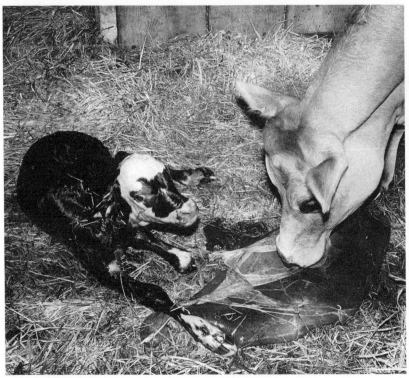

(Below) *Cow begins to clean up, and prepares for a surprise second calf.*

- The waters have burst and the cow has been straining *intensely* for a couple of hours with no sign of the calf: *get the vet.*
- A properly presented calf seems to be stuck at the hips: *get the vet.*
- An apparently properly presented calf, front feet first, has not emerged further after about eight hours: *get the vet* for a possible malpresentation.

Malpresentations

You *must* seek experienced help if there is a malpresented calf. It is not as easy to untangle a calf in the womb as it is a lamb and you are bound to make matters worse. The proper presentation is both front feet first, with the nose resting along the front legs as if the calf is preparing to dive out into the world. Suspect a problem if there is only one foot showing, or if the feet emerge soles uppermost (they are probably hind feet, or, if they are forefeet, the calf itself is upsidedown), or if the first thing you see is the calf's back end (breech presentation).

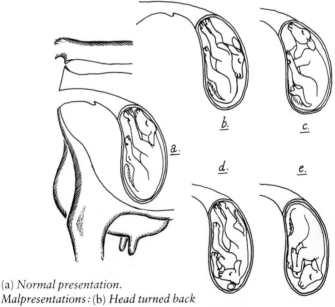

(a) *Normal presentation.*
Malpresentations: (b) *Head turned back*
(c) *Forelegs back* (d) *Hindlegs first* (e) *Breech position.*

After the birth, the cow should automatically clean up the calf but an inexperienced heifer might want nothing to do with the thing and you should wipe away the membranes with clean straw, especially from the nostrils so that it can breathe. If a calf seems too weak to try, help it take a first breath by carefully tickling a straw inside its nostrils. In a real emergency, shock the breath out of it by throwing a pailful of cold water over its head and chest or give artificial respiration if you know what you are doing. *Dip the residual umbilical cord* in iodine to deter infection. It will gradually shrivel and drop off of its own accord a few days later.

On the whole, and in principle, *leave the calf and cow alone* to establish the maternal bond with lots of licking, sniffing, and that delightful, special, intimate tone of voice a cow reserves entirely for her own calf. Enjoy it; and watch discreetly to see that, eventually, the calf manages to locate a teat, after plenty of misses and false tries, and takes its first drink, often within an hour of its birth. Do not interfere unless it has failed to suckle within the first five or six hours — a crucial period during which it *must* get its first meal, not so much to fill its belly as to absorb vital antibodies that enable it to fight the dangerously infectious environment of this new world beyond the womb. After its first suckling, a calf passes **black, jelly-like foetal droppings**. This is the evidence you need that it has suckled if you were not there to see it drink.

THE CALF

The choices for calf-rearing are:

Natural rearing
- By its own mother, either running with her or kept separate and allowed to suckle at intervals.
- By a foster cow, likewise running with her or given limited access for suckling, either as a single calf, or one of two or more on the same cow at one time or in succession.

Artificial rearing
- Bottle-fed, either with fresh cow's milk or with reconstituted calf-milk powder.
- 'Calfeteria' help-yourself bottle-feeding system with re-constituted milk (usually cold).
- Bucket-reared, i.e. sipping rather than suckling.

Calves naturally do best suckling cows, preferably their own mothers, and preferably running freely with them because there is so much more to life than food; there is a lot to learn about, for a start, and also the physical presence and attentions of the mother pass on certain immunities to the calf. In addition, by running out in the fields it has every opportunity of keeping fit, well exercised, disease-free, unstressed and benefiting from sunshine. Suckled calves grow more strongly *and* more cheaply than those which are artificially reared. Take the example of internal parasites. It is noticeable that calves running in the field with their mothers are far less likely to suffer badly from worms, even if there is something of a worm-burden in the field, whereas an artificially reared calf in the same field would be pulled down and lose condition in no time.

As soon as a calf is removed from its own mother, the battle begins and the calf is usually on the losing side, especially against disease, partly because the stress of separation reduces any animal's resistance and makes it more vulnerable to infection, digestive upset and so on.

As soon as an animal of any age is housed, the fact of confinement in itself heightens the risk of disease. It is very difficult to keep a building absolutely clean, and anyway housed cattle are in close contact with each other so that infections are more readily passed between them. There are those who argue that an outdoor environment is harsher but it remains a fact that suckler herds running on the hills in some of Britain's worst climates are far hardier and healthier than dairy cows given every comfort. It can be argued that this is partly because of the excessive production demands made on dairy cows and on a genetic loss of hardiness precisely because of being bred to meet those demands, but I have acquired cows from commercial dairy herds and changed their life-styles radically without changing their genetic structure and, time and again, the outdoor cow has remained far healthier than she ever was on the dairy farm.

Calves are tough little things. It is rarely the weather that brings them to grief but frequently infections picked up indoors. If circumstances permit, then, let your new calf be reared by its own mother and let it run with her. Let it keep company with her most of the time even if you want some milk for yourself. Put a motherless calf on to a willing foster mother, rather than use any artificial rearing system or, as a much less satisfactory compromise, rear it on a bottle. Bucket-rearing, though thought to be easier for the human, is by far the last choice for the calf.

Jersey calf lying up safely in undergrowth.

Colostrum

There are three crucial stages in the feeding life of a calf: the **first four days** (when it needs colostrum), the **first three weeks** (while its stomachs are developing), and the **time of weaning**.

Whatever its future, every newborn calf needs as a matter of urgency its full share of colostrum. **Colostrum** is a thick, sticky, yellow first-milk produced by the freshly calved cow for the first three or four days after the birth; thereafter her milk becomes normal. Colostrum is highly digestible to the calf and about 40 per cent more nutritious than ordinary milk. It contains concentrated vitamins, including five times as much carotenoid (precursor of vitamin A); it contains twice as much calcium, four times as much protein and more than twice as much non-fat solids. Above all, it contains about sixty times as much gamma globulins — simple proteins which play a vital role in transferring passive immunity against disease from the cow to the calf. Note that the cow also passes immunity (to diseases familiar in her own environment) by close contact with the calf. Without colostrum, however, a calf is unlikely to survive long enough to acquire that further immunity.

A **newborn calf** needs:

- About one pint of colostrum to avoid death from blood-poisoning.
- About six pints of colostrum to avoid diarrhoea.

If for whatever reason a cow cannot suckle her newborn calf, you must take steps to ensure that it does get its ration of colostrum — preferably its own mother's, whether fresh or frozen. The next best substitute is another cow's colostrum (keep an emergency supply in the freezer). If there is nothing else, goat's colostrum might help but not a lot, and in a dire emergency you could make up an artificial substitute though it won't help a calf whose environment is less than healthy.

> **Emergency colostrum substitute:**
> *1 fresh egg*
> *0.75 litres (1½ pints) of cow's milk*
> *0.25 litres (½ pint) of warm water*
> *1 teaspoonful of cod-liver oil*
> *1 dessertspoonful of castor oil*

Beat the egg into the milk-and-water, add the cod-liver oil, and bottle-feed three times a day, a litre at a time, for four days. Include the castor oil as a laxative until the foetal dung has been passed.

The digestive system

Many artifically reared calves have digestive problems — indeed these are probably a major cause of regression and death among batches of housed calves. Scouring (diarrhoea) can be a killer in that it deprives the calf of nourishment and leads to dehydration and general weakness, so that the calf is immediately more susceptible to other problems. Bucket-feeding is the worst offender, as shall be explained.

The digestive system of a newborn calf is not the same as that of a cow. Its **rumen** is undeveloped, so that the calf is incapable of dealing with roughage. In contrast, the **abomasum** is three times the size of the rumen and milk is channelled straight into it

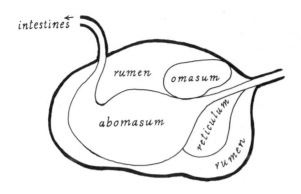

The calf's digestive system.

as the calf suckles, by-passing other stomachs. Milk is the only food from which a new calf can obtain sustenance.

As the calf suckles, the milk forms small clots in the abomasum and offers maximum surface area for the work of digestive juices. Pause now to think about bucket-feeding. For a start, a suckling calf's natural posture is with its muzzle lifted up to the udder, stretching its throat so that the milk can pass down the special oesophagal channel into the abomasum. A bucket-fed calf has its head down instead of up, and it has to learn to lap the milk rather than suck it, probably drawing in air as it does so (giving it indigestion), and probably (once it learns how to drink) wolfing it down in large draughts so that it forms one or two big indigestible lumps rather than lots of small clots. The second problem is that udder-milk is body-warm and, too often, bucket milk is not only artificial milk but is often fed too cool or even cold.

By about the second week of life, a calf is vaguely toying with wisps of fodder. In due course a little is swallowed and this begins to stimulate the gradual development of the rumen. At three weeks old it is *theoretically* possible for a calf to digest small amounts of solid food but it gets far more nutrition from milk. At eight weeks old its stomach proportions have reversed: the rumen has become about three times the size of the abomasum and the relative importance of milk and solid foods are likewise gradually changing around; as the calf grows it eats more food rather than demanding a higher level of milk output by its mother, though milk continues to be important for almost as long as the cow is prepared to supply it. However, the fast-growing calf gradually eats and digests more and more fodder.

- *A very young calf depends on milk* and does not have the stomach for solid foods, be they fodder or concentrates.
- *A calf must have access to fodder fibre* in order to be able to develop its rumen, gradually. Fibre is much more important for its proper development at this stage than other solids.

A calf born in the spring has the benefit and pleasure of growing with the grass; and the cow, too, can replenish her resources, depleted by the suckling calf. If you are happy enough to let the calf (or calves) have all the milk, leave them to it, but when the calf is voluntarily eating solid food let it have private ad lib access to extra food if necessary, especially if the grazing is not at its best. Provide it with the best-quality hay, crushed cereals, flaked maize and so on. **Creep feeding** is an arrangement to let a calf have access to extra food while denying it to the cow. Make an area which the calf can enter easily but the cow cannot; have a calf-wide gap between two posts through a fence or into an enclosure, or put a horizontal rail across the gap high enough for the calf (remembering that it is growing) but too low for the cow to pass under.

Fostering

To encourage a cow to accept a foster calf
- Choose a cow who actually *likes* fostering. Some do, some don't, and there is little point in forcing the situation on an unwilling cow.
- Confine her at first in a loose-box. Tie her by the halter and give her something to eat as a distraction. Be sure that the new calf actually knows how to suck (bought-in calves might have been bucket-fed). Let it into the box and let them sort it out for themselves. Watch to see that the foster calf does eventually get some milk but be ready to bottle-feed if really necessary.

Unlike some species of mammal, cows are not auntie-figures and might not tolerate another cow's calf; they consider their duty is entirely to their own offspring. There are a few tricks whereby you can try to fool a cow into thinking the foster actually is her own calf but it's harder than fooling a ewe. Your first weapon is smell; try disguising the calf's alien smell by rubbing it with something even more alien like eau de cologne! The cow might be so confused that she won't notice the calf. In theory. You could also try a blindfold, in that most animals are more quiescent when they are in the dark. Some people try brute force, constraining the cow in such a way that she can't kick the calf away, but any cow is capable of withholding her milk if she is really determined and, remembering the principles of this book, you should be on her side, not fighting her.

Weaning

Suckler calves on extensive beef systems are usually weaned at about six months old. In the wild a calf would happily go on taking a swig until the cow's supply dries off naturally in the later stages of pregnancy (say, a month before calving) or until she gets fed up with being bullied and kicks her large offspring away from the udder. She will certainly do one or the other in preparation for her next calf; the older one really does not *need* milk any more.

Milk is an excellent food and you should let the calf make the most of it — it's the best nutrition for growing a good beefy calf — but suckling inevitably becomes a drain on the mother and you should be guided by *her* condition as much as by the calf's to decide when it should be weaned, even if the mother is still willing to suckle. In any event a cow deserves a good long rest in preparation for her next lactation and you should certainly wean the calf at ten or eleven months old at the latest. The rest is even more important if she has been asked to raise more than one calf, whether two at a time or several in succession.

Calves and milking

What if you want some of the cow's milk yourself? This is sometimes a dilemma, or even a practical problem. In India and Africa, even the most well bred indigenous zebu cows refuse to let down their milk for humans unless their calves are at least present and, preferably, already suckling. In ancient Egypt about 4,000 years ago, a calf was tied to its mother's foreleg at milking to stimulate let-down while keeping the calf away from the udder.

British cows have a long history as dairy cows deprived of their calves and have been selectively bred for ready let-down anyway, but that will not deter a strong-minded cow who does not really see why you should have a single drop of her milk instead of the calf, even if there is more than enough to spare. She might even fool you into believing that she hasn't *got* any milk and then, the moment your back is turned, her gleeful calf is hard at work at the udder and the stuff is positively pumping out! Or it might be that the calf has suckled shortly before you come along — and from bitter beginner's experience I can promise you that the calf has *always* just emptied the udder, whatever unexpected time of the day or night you try to take them by surprise. An average British cow of a dairy breed (Jersey, Guernsey, Friesian, Ayrshire and so on) yields enough milk every day for perhaps three of four calves at once: there is *plenty* for one calf to share with you.

There are perhaps two reasonable solutions. One is to have at least two cows in milk at the same time and to persuade one of them to foster the other's calf so that you can have all the second cow's milk yourself; or to have two cows calving about six months apart so that the first cow's calf can be weaned at a few months old and you can then milk the cow until the end of her lactation, by which time (with luck) the second cow's calf can be weaned and you can swing your milking pail under her instead. The second is to limit a calf's access to its mother's udder and then either let it suckle a little before or after you take over or let it suckle from one side (most calves have a preferred side) while you milk from the other. This compromise should encourage most cows to be co-operative.

However, bearing in mind the basic principles of this book, you should not follow the common practice of entirely depriving the calf of its mother's company between suckling sessions. Many people keep the calf indoors in a loose-box, letting the cow in at intervals to suckle and be milked, and that is efficient but not very fulfilling for the calf except for the first week or two of life. Remember the wild calf, quite happy to 'lie up' alone all day except for when its mother returns to suckle and groom it every few hours; but during its second week the calf will probably join the herd, and its lying up will be in the company of other calves by day and with a group of mothers by night.

Even if the housed calf shares its loose-box with other calves, it is being deprived of outdoor experiences and the whole business of learning about life from older animals. A possible compromise, then, is to accommodate the calf in the field with the cow (or within her winter quarters) in such a way that the two can communicate with each other visually, audibly and tactilely but the calf cannot actually reach the udder until you let it join you during milking.

At its simplest, you could build a field shelter with a section for the calf partitioned off with, say, strong vertical poles set close enough together to prevent the calf from pushing its muzzle through but wide enough for it to see out, and high enough to confine the calf but low enough for the cow to reach her head over for grooming and general conversation. The calf can see what is going on in the rest of the field and can make physical contact with its mother, except for her udder. But do not keep the calf confined all the time: it needs the chance to rush around the place, investigate things, sample plants, chase birds and butterflies, feel the sun and the rain on its back, learn and have fun and mix with other animals and all the other joys of calfhood which can give so much pleasure to the watching cow-keeper.

If the calf has the company of other calves in the field, life should be easier for the frustrated milker. Bear in mind those wild calves, very soon forming their own little group within the herd and sleeping off their meals in a voluntary 'crèche' while the mothers graze. If you have two or three calves, try giving them their own little paddock within the cows' field so that you can milk a mother when you choose before letting the calves have access. That way they have company and healthy outdoor stimulation while you control the milk supply.

Every circumstance is different, of course, and it will not take much ingenuity for you to work out your own system for sharing the udder, though I do beg you to consider the fuller life of both cow and calf as well as your own needs.

Some cows are actually quite relieved not to have a calf constantly butting at the udder and demanding a feed. You should anyway keep a check on the udder, even if you are not milking, to make sure there is no sign of mastitis (see p.105). Some calves become very rough, especially as they grow, and if a cow is persistently kicking away her own calf it could be that she is very sore or her teats have been damaged by a calf's teeth.

Artificial calf-rearing

If you choose to rear calves by hand, remember that you are effectively dealing with orphans. It is not being sentimentally anthropomorphic to understand that an orphan animal is under a great deal of stress even before humans decide to take charge of the rest of its life. Act with firm sympathy at all times: you have a helplessly dependent creature in your care and it was your choice, not the orphan's.

The wild herd's mothers tend to group together overnight with their calves. They rise at dawn; the calves suckle and retire to their crèche while the cows graze. A few hours later they begin to call for another meal; a couple of mothers respond and soon the rest of the cows follow suit. There will usually be another mass-suckling in the afternoon and again in the evening before the cows and their calves group together again for the night.

An artificially reared calf would welcome a similar feeding pattern but, for human convenience, it usually has to learn to drink less often and in larger quantities, which is not good for its stomach. However, it will no doubt accept the routine in due course — and it must *be* a routine, with absolutely regular feeding times, and with a consistency to the make-up and temperature of the feed. Any change should always be introduced gradually.

A very young calf, remember, would in the wild be lying up for the first week or two and is therefore best kept in its own private hideaway, protected from draughts but with adequate fresh air and no artificial heating unless it is ill. Make a straw igloo for it (high enough so that it can stand), using a few bales for walls. When it is a week or two old, let it join a very small group of calves of a similar age in a well-strawed loose-box — do not follow the typical commercial practice of restricting such calves within individual pens which, though it makes for easier management at feeding time, deprives the little ones of normal contact. If your calves and their environment are healthy, and if their urge to suck is satisfied by the feeding method, there is no need to keep them separate. Bucket-fed calves, however, without the satisfaction of suckling, are often desperate to suck *anything* and have to be kept apart from each other to deter them from molesting each other's navels, ears and anything else they can channel their tongues around.

How often?
Ideally be guided by the wild pattern: feed three or four times a day (dawn, mid-morning, mid-afternoon and late evening). *Be strict with time-keeping.*

How much?
The basic formula is: 1 weight-unit of milk for every 10 units of the calf's bodyweight, i.e. 1 litre per 10kg, or about 1 pint per 15 pounds.

    45kg Friesian: 4.5 litres/day (1 gallon)
    30kg Jersey: 3 litres/day (5 pints)
    35kg Dexter: 3.5 litres/day (6 pints)

Take special care if you are using powdered milk substitute (specially formulated for calves), and follow the rules *every* time — it takes just one slapdash mix-in-a-hurry feed to wreck a calf's digestion and set it scouring (diarrhoea).

- *Be scrupulously clean:* sterilize all feeding equipment every time it is used, including the mixing equipment.
- *Mix properly:* follow the instructions about the correct proportions of powder and water exactly, and mix in the powder very thoroughly with no lumps at all.
- *Be consistent:* feed at the same temperature every time, preferably warm rather than cold; and feed at the same strength unless it is necessary to weaken it because of scouring.
- *Keep it fresh:* mix a new batch for each feed, every time, and only mix it just before feeding.

To satisfy the sucking urge, which also ensures the milk is channelled into the abomasum and forms the little clots for easier digestion, use either a hand-held bottle for each calf (with a calf-size teat and an appropriate size of opening) or a help-yourself system which can be adapted to feed several calves at once — but make sure each gets it share. The basic calfeteria is a bucket with teats protruding from holes near the base and there are several manufacturers' designs which are variations on this theme. The main drawback is that, if it is used on a suckle-when-you-wish basis, the milk will be fed cold and is sometimes left lying around too long so that it is less than fresh.

Make sure that the very best hay and fresh drinking water are always available, even if the calf seems uninterested in them. As already explained, it is important for top-quality roughage to be eaten in very small quantities at first to help develop the rumen properly, and you should not even think of feeding cereals or pelleted concentrates (if at all) until you are sure the calf is already digesting fibre in hay, soft straw or grass.

Calves rarely need to be taught how to suckle, whether the teat is real or rubber, but they do need to be taught how to drink from a bucket and it can be quite a challenge for both of you (another point against bucket-rearing).

To train a calf to the bucket

- Exercise great patience and kindness — never rush, never lose your temper. Do not train a calf unless you have the right temperament.
- Back the calf into a corner to confine it.
- Squat in front of it with the bucket of milk between your knees.
- Offer the calf your clean fingers, which it will eagerly suck.
- Lower your fingers gently into the milk so that the calf's muzzle just touches its surface — take care not to let it snort milk up through its nostrils.
- While it is still sucking your fingers, gradually withdraw them from its mouth. With luck, it will find itself sucking milk, but it is an unnatural act for a young calf and might need several finger-dipping sessions.
- Take great care that it does not gulp the milk down and choke itself, nor drink it too quickly once it knows how.

5. Good Health

Animals are intrinsically healthy. Herbivores such as cattle are usually even healthier than carnivores in the wild and normally remain so until they are killed by predators, by accident, by severe territorial fighting, by famine, by old age or, very occasionally, by epidemic disease. Illness, as such, is not a common occurrence.

A domestic animal has all aspects of its environment controlled by humans, and that is when its problems begin to germinate and fester. The very fact of confinement, though it has great advantages to livestock owners and a few advantages to the livestock as well, immediately exposes the animal to possible ill-health from a combination of stress and disease.

WELFARE

If your own animal management 'system' is firmly based on concern for the animal's welfare rather than on its economic potential, you are off to a good start. Too often, though, ignorance of animals' needs blunts the act of caring or, worse, the priorities are reversed: the animal is seen as a productive machine, a servant rather than a partner, a creature to be controlled rather than accommodated, a prisoner rather than a guest, and in such circumstances it is hardly surprising that so many farmers fight a constant battle against livestock disease, frequently relying on blanket prophylactic drugs rather than good husbandry to prevent illness and on conventional drug-based treatments which treat the symptoms rather than tackle the underlying conditions. They would do better to come to terms with why there are problems and to deal with them by sound basic husbandry rather than by lining the laden pockets of veterinary pharmaceutical companies. Drugs should never be used routinely but only when they really are the best option.

An animal with the best possible chance of remaining healthy is:

- Treated with due respect based on a proper understanding of its nature.
- Given a spacious and healthy environment.
- Allowed to express its natural behaviour as much as possible.
- Given a balanced, clean diet appropriate to its species and a constant supply of fresh, uncontaminated drinking-water.
- Relieved of the burden of stress caused by excessive production demands.
- Bred from genetically sound stock so that it is not inherently prone to disease or deformity.

With the animal's welfare as your prime consideration, you have a duty to find a good veterinary surgeon who understands animals as individuals and who has sympathy with your principles, whatever your views on the veterinary profession as a whole and whatever your principles concerning animal health. However well meaning, you are not a professional in the matter of *diagnosing* a problem; there could be something far worse underlying whatever it is that you recognize and treat. Find somebody you can trust who has a thorough training in basic animal physiology and preferably somebody who is not only qualified in conventional veterinary medicine but who also keeps an open mind and has actively investigated and understood the potential value of complementary practices such as homoeopathy, herbalism and acupuncture. For example, the Oxfordshire vet Christopher Day is a pioneer in the field of homoeopathy on the farm and has already attracted (and kept!) several regular clients who, though originally sceptical commercial farmers, find that his methods actually work, on a herd basis as well as for individual animals. Read everything he has ever written on the subject.

In an organic farming system, a compromise is made over the use of veterinary pharmaceuticals. The routine use of prophylactic drugs (preventive medicine) is avoided, though vaccines, intended to protect animals against infection, can be given if there is a *known* and specific disease problem on the holding, bearing in mind that any such vaccination can interfere with the *natural* immune system and have a detrimental effect on it as a whole, as well as potential unpleasant side-effects. It is much better to deal with the environment that is harbouring the disease and to build up the animal's natural immune system than to blast the creature with vaccines without removing the cause. Likewise, the Soil Association does not recommend the routine use of anthelmintics to control internal worms: it is much more important to break the life-cycles of typical parasites by management of the grazing and general good housekeeping, only using anthelmintics, if at all, when an individual animal has become infected. The same is true of external parasites: they can often be controlled by manipulation of the environment and mechanical rather than pharmaceutical treatment, which is practicable on a small scale although laborious for a sizeable herd and even worse with sheep, who seem to attract some of the most repulsive and persistent parasites.

Environment is so often the key to continued good health. You know how it is yourself: you can be a thoroughly healthy person in spite of facing appalling weather conditions as you struggle outside to feed and water the animals, with never a day's illness until you find yourself confined with other people in a heated place — the dentist's waiting-room, a train or plane, an office or shop or school — and then, bingo!, you catch a cold or 'flu or worse. The enclosed environment, especially if it is artificially heated, fosters infection, and the same is true for livestock.

General condition

Quite apart from how well a cow feels, keep an eye on her general condition to see that her diet is adequate and that not too much is being demanded of her. Don't be deceived by Jerseys in particular: they are naturally on the ribby side with a rather

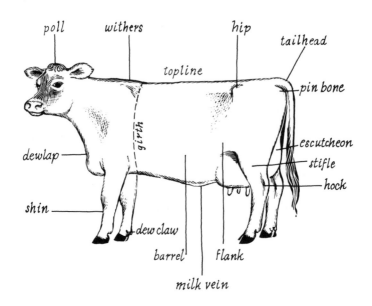

bony look even in show-ring condition, and an alarming hollow before the hip bone, but they should retain a good 'bloom' of health.

If you cannot trust your eye (and looks can deceive), 'weigh' the cow regularly to be alert to changing trends. That does not mean literally plonking her on the bathroom scales or even finding the nearest weighbridge: use a girth-tape — you can buy (or devise) tape-measures which are marked in approximate weights as well as chest-girths, as it is generally reckoned that chest measurements accurately reflect weight (see diagram).

Weighband Approximations:

```
100kg (220lb)  = 104cm (40in) girth
200kg (440lb)  = 133cm (52in) girth
300kg (660lb)  = 153cm (60in) girth
400kg (880lb)  = 170cm (66in) girth
500kg (1100lb) = 185cm (72in) girth
600kg (1300lb) = 196cm (77in) girth
```

Hoof and horn

The condition of an animal's feet makes much more difference to the general state of health than might be appreciated. Hoof-horn, like human fingernail, is constantly growing, renewing itself as the surface is eroded by daily life, but sometimes, indeed quite often, it grows faster than it is worn away, especially at the front of the foot. The result of overgrowth is that the shape of the hoof alters and the whole balance of the cow is distorted, which means pressure is put on all sorts of unlikely parts of the body as they become misaligned, especially spine, shoulders and hips. All for the want of a toenail trim! See to your cow's feet *before* she develops inflamed joints; trim them anyway twice a year and do your best to give her different surface textures to walk on and stand on, with regular walks on concrete as well as on pasture or straw in order to wear them down. (If you are working your cow as well, you'll almost need to have her shod.)

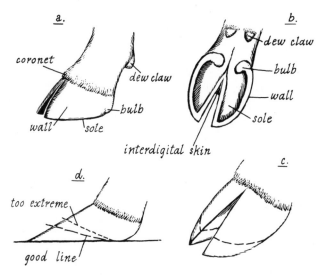

To trim a hoof

- *Watch an expert at work* until you are really competent to do the job yourself — talk to your vet or a local farrier, or an itinerant hoof-trimmer, and read Toussaint Raven's excellent book on the subject (see Bibliography).

- Use blacksmith's pincers or a strong, sharp knife.
- Aim to trim the hoof to its correct shape so that the cow stands in balance; the flat base of the hoof should be at a right-angle to the line of the shin bone.
- *Be very careful indeed* not to cut the quick, nor to pare away so much that the thinly protected undersole can become bruised.

Feet can also suffer from infections, usually after some kind of mechanical injury, and especially between the claws. If there is the slightest suspicion of lameness, take a closer look at the favoured foot and sort out the problem — perhaps a small stone or nail or wire-barb has entered and needs removing. Certain managerial practices lead to lameness as a metabolic disorder making the feet hot and tender.

There are some standard techniques which are nothing to do with maintaining or repairing health but which are practised by many livestock owners for management purposes. They are all what you might call mutilations — the removal of horns and testicles being the most common farm 'operations'. Most have no place in this book but the question of **horns** is less easily dismissed. Some breeds of cattle are naturally polled — they are genetically incapable of growing horns — but all the other breeds are naturally horned, except for the occasional genetic mutation. The horns vary in different breeds from little short things to big dramatic sweeps — though the British breeds cannot compare with some of the amazing horns in African cattle.

It is a matter of personal taste but I feel that horns set off an animal properly and look more balanced and 'right'. However, you rarely see horned animals, particularly in commercial dairy herds, and for good reason: horns in confined spaces can cause damage to other animals and to people and, if they happen to be snapped off in an argument with an immovable object, the breakage causes considerable pain and a positive fountain of blood. They can also grow awkwardly and the tips might need to be filed down or trimmed off with cheesewire to prevent them from literally growing into the animal's cheek, but be extremely careful not to go near the quick, which is the living part of the horn, packed with nerves.

It is standard practice in most dairy herds and many beef herds to deprive young animals of their ability to grow their horns. This is by the technique of **disbudding**, which is the act of killing the potential growth of a calf's horns when they are no more than little buds just beginning to burgeon from a calf's crown.

Weight the pros and cons of disbudding carefully — it is an irrevocable act and inevitably causes a certain amount of at least discomfort and immediate stress to the calf, however carefully done. In the wild, horns serve a purpose, largely as defence against predation but also to establish 'hook' orders in the herd hierarchy and, when they are long enough, as a very useful aid to scratching hard-to-reach parts of the body! The horns of some African cattle, according to 18th-century travellers' tales, were frequently trained into elaborate shapes and directions, or even made to grow together as unicorns or divided to grow as multiples. Ouch!

Calves are disbudded at about 2-3 weeks old, when the two buds can definitely be recognized. Some of the traditional techniques were thoroughly unpleasant and potentially dangerous but the modern method is to use a special hot iron applied briefly to each bud after giving a local anaesthetic. *Do not* do it yourself: the anaesthetic needle must be very precisely placed. Talk to your vet. If a 'friend' offers to do it for you, don't be shy about insisting that the needle is properly sterilized and sharp. I have seen some fairly horrific practices on commercial dairy farms and the results of using dirty or blunt needles for the anaesthetic are as nasty as the bungling of the hot-iron technique. Sadly some farmers, dealing with large numbers of animals over the years, become not only casual but casually brutal as well, though they are the exception rather than the norm.

Castration is one of those unpleasant routines to which many farmers inevitably become inured, their sensitivities dulled by numbers. When you keep a male calf for beef, you can in theory leave it 'entire' if you bear in mind two main problems: it will become increasingly obstreperous as it matures, and it could cause an unwanted pregnancy. It is possible, in theory, to 'work' a bull calf as young as 10 months (though never in practice) — some breeds are naturally precocious and some suckler-herd owners separate cows with bull calves from cows with heifer calves when the youngsters are six months old, just in case. The cows would not tolerate such immature advances but the heifers might have little option. Normally a bull would not be deliberately used for service until he is at least 18 months old, and his semen production does not reach adult levels until he is about three years old.

The traditional methods of castration are by knife, by rubber band and by so-called bloodless castrators which are like heavy pincers for crushing the spermatic cords. Ouch — very much ouch, whatever you use, and however much people tell you it doesn't really hurt the animal, it does.

By law, only a vet can castrate a male calf more than two months old. By law, too, rubber bands (applied with the aid of an 'elastrator') can only be used within the first week of life, but they cause much distress and often lead to sepsis. Don't even think of using a knife, however sharp and sterile, unless you know exactly what you are doing: ask a vet rather than a farmer to train you. The bloodless or 'Burdizzo' castrator is theoretically easier to use and less likely to cause infection as there is no open wound but, again, you need to be experienced and to work initially under proper supervision.

Techniques

There are certain mechanical techniques that every cow-keeper is likely to use at some stage and in which it helps to be proficient in advance, so that there is less uncertainty in the face of reality. They include routine maintenance such as foot-care and horn-care, and nursing procedures such as giving pills, drenches and injections. The last two are ways of administering treatments rapidly so that they are quickly effective and they can be life-savers in an emergency. Doses can also be given in food (or drinking-water) and it helps to disguise them with the smell of aniseed or in molasses, which most cows like.

A drench is a liquid administered orally, usually medicinal or to counteract dehydration or malnutrition. To give a drench:
– Stand by the cow's head, facing forwards.
– Put one arm over her to hold her jaw on the far side: insert your fingers in the gap by her back teeth.
– Gently raise her muzzle so that her neck and head are slightly stretched (to open a clear channel along the gullet) a little above horizontal, but keep the mouth below eye-level.

– Using a bottle with its neck protected from accidental damage (a piece of rubber tubing), insert its neck into the cow's mouth.
– *Very* carefully and slowly trickle the liquid down her throat. You want it to go straight to her stomach, not into her lungs.

To take temperature:
- Lubricate a clean veterinary thermometer with vaseline and take the temperature rectally.

To give injections:
First take practical lessons from a vet.
Always use a new, sterilized, sharp needle.
Always clean the surrounding skin beforehand.
Massage the site afterwards to avoid lumps.

- **Subcutaneous** injections are given just under the skin: pluck up a loose fold of skin behind the shoulder and carefully slide the point of the needle into the hollow you have created under the hide.
- **Intramuscular** injections are given into a muscle, usually on the thigh or behind the shoulder. Deaden the target area with a quick thump and then boldly drive the needle straight into the muscle, avoiding bones and blood vessels.
- **Intravenous** injections are given into the jugular and should *never* be given by amateurs.
- **Intramammary** injections are sometimes used against udder infections. No needle is involved: you simply insert the nozzle of a special tube of ointment into the teat orifice and squeeze the contents up the canal.

EMERGENCY FIRST AID

While waiting for professional help:

– *Reassure* the animal.
– *Keep it calm*, quiet and warm in a private place with subdued lighting.
– *Do not move* a shocked animal or one which might have internal bleeding or spinal damage.

Bleeding: Stop by applying pressure — from above the wound if an artery is bleeding (it will be spurting) or from below for a vein (it will be flowing). Use fingers or padding.

Breakages: Don't move unless essential but give reassurance. A calf with a broken limb can be carried in the usual way with the legs hanging down freely.

Bruising: Use vinegar or tincture of iodine.

Burns: Reduce the heat as a priority: immerse the flesh in cold water for 10-15 minutes. Do *not* use lotions or oils to soothe but some strong tea provides tannic acid to form a protective scab more quickly. Keep very clean; cover with clean, dry bandage when dry.

Electrocution: *Switch off the power supply!* If you can't, use a *wooden* walking-stick to draw the animal away from contact with a live wire. Give resuscitation if not breathing, or treat for shock (see below).

Eyes: Use saline wash or put a dab of eye ointment on your finger tip to remove a foreign body.

Fever: Use veterinary homoeopathic aconite (in either water or food) as the first treatment in early stages if you are familiar with homoeopathic principles.

Resuscitation: If unconscious and has stopped breathing, kneel beside the animal, spread hands over ribs just behind shoulder blade, keep arms straight and rock forwards to apply firm pressure; promptly rock back again to allow lungs to fill with air. Repeat after about six seconds, and continue at similar intervals as necessary.

Shock: Symptoms include abrupt fall in blood pressure, cold extremities (but not shivering), fast weak pulse, rapid shallow breathing, pale membranes. Keep the animal quiet, warm and undisturbed; do *not* move it. Use homoeopathic arnica as the primary treatment (given in water or in food), whatever the cause of the shock or stress.

Stings: Watch out for stings in mouth or throat, which could swell and obstruct breathing (seek veterinary help). Soothe stings with ammonia, bicarbonate of soda, or cold water, or rub with half an onion.

Wounds: Clean surface with saline solution only (1 tspn salt to 1 pint water) until vet has checked the wound. Let cow's own saliva cleanse superficial grazes and scratches; black pepper is a useful standby disinfectant too. Witch hazel is a good, safe astringent and cobwebs really are recommended for staunching bleeding.

SICKENING

If you know your animal well, you will spot that she is off-colour almost before she knows it herself, just as you would with a person who is close to you — it's almost telepathic. A good stockman will perceive instantly if any animal, of whatever species and whether a familiar individual or not, is not well. It is what they call the 'stockman's eye' but it is almost a sixth sense; it comes partly from experience but largely from an empathy with animals. It's why good mothers make good stockmen!

A sick cow:
- Stands apart from the herd, head held low, eyes dull, disinterested in her surroundings, or
- Has a fixed, staring look, ears held back, nostrils flared (probably in pain).
- Fails to groom herself and probably has a 'staring' coat.
- Is disinterested in eating, and disinclined to cud.
- Might smell wrong — breath, milk or dung.
- Might have very loose dung or be constipated.
- Might kick her own belly or grind her teeth — she's in pain or discomfort.

A cow under great stress has constant, rapid eye movements even though she might be standing stock-still.

NORMALITY
Normal temperature: 38-39°C (101.3°-103.1°F)
Normal pulse (tail base): 45-60 beats per minute
Normal respiration rate: 12-16 per minute

Nursing

A sick cow needs a quiet, private environment in a dry, draught-free, comfortable loose-box. It is one of the few times when this sociable animal actually prefers rather than fears isolation.

Herbs for health

This is such a substantial subject that it can only be touched on here: invest in a really sound animal herbal if you are interested. Let animals be their own herbalists; give them access to a wide range of herbs in the field, growing in pasture or deliberately planted on the headlands and in the hedges.

Note, however, that several plants which are excellent as tonics or medicines in small amounts can have adverse effects in larger 'doses' or could even be poisonous. Treat herbs with great respect and also appreciate that different herbs suit different individuals. Here are some of the most useful.

MEDICINAL HERBS

COMFREY (high-yielding plant, full of protein) is also known as knitbone . . .

GARLIC — many applications, including vermifuge, mastitis cure.

Other parasite expellents include:

Mustard	Couch grass	Brambles	Rue
Wild turnip	Elder browsing	Broom tips	Vervain

Pregnant cows benefit from:	**Freshly calved cows** benefit from:
Wild raspberry leaves	Crushed nettles
Blackberry leaves	Watercress
Wild briar leaves	Watermint

General tonics (let cows choose for themselves):

Hawthorn	Elder	Ivy (a little)	Cranesbill
Hazel	Willow	Herb robert	Ground ivy (dry cows)

Poisoning

The main agents of cattle poisoning are lead paint in old buildings, discarded chemicals of various kinds, and certain plants. Calves also eat all sorts of unsuitable things like plastic, twine, balloons and anything else they try for the experience, but these are usually indigestible rather than poisonous, though they can cause severe problems by becoming tangled up internally. The best cure for all such problems is prevention by vigilance; check the environment regularly for hazards.

POISONOUS PLANTS

Poisonous green or dried
Belladonna family   Bracken   Buckthorn   Cowbane   Flax   Foxglove
Horsetails   Lily-of-the-valley   Lupin   Ragwort   Yew

Poisonous in quantity *(some are medicinal in small amounts)*
Rhododendron   privet   potato/tomato stems/leaves   acorns
ash-keys   laburnum (all parts of the plant)   box   mistletoe berries
spindle   ivy   bulbs, corms, rhizomes   buckwheat, broomrape,
charlock, white mustard, wild radish, horse radish, hemlock,
houndstongue, hypericum, meadow saffron, hard rushes, sedum, sorrel,
spurges and mercuries, tobacco, water figwort

If you do suspect an animal has consumed something
poisonous, you need to take very fast action indeed. Contact the
vet immediately and, while waiting, take some first-aid steps if
you know exactly what the poisonous substance was. The aim in
most cases is to get the stuff out of the animal's system or
neutralize it before it can cause irreparable damage, and to
mollify any corrosive effects it might have had already. The
victim will need plenty of good nursing once the poison has been
removed but some animals never learn; they seem to develop a
craving for the very stuff that poisoned them and you must in
future make absolutely certain it is never within their reach.
Antidotal action (for extreme emergencies only — a vet will have
correct antidotes for specific poisons):

- **Acids:** Neutralize with alkaline substance (e.g. bicarbonate of
  soda in milk).
- **Alkalis** (ammonia etc): Neutralize with dilute acids (e.g.
  weak vinegar).
- **Lead** (flaking paint, tarpaulins etc): Violent symptoms. Use
  Epsom salts, milk, egg whites or strong tea but get urgent
  help.
- **Yew, laurel**: Sugar, or ammonia fumes.

To soothe membranes (mouth and stomach) after irritant has
been dealt with: 12 eggs and 1 lb sugar, in half a gallon of milk.

Diarrhoea is initially *good* — helping to flush poisons out of the
system — but if persistent it will cause severe dehydration. Give
kaolin or chlorodyne, or boiled milk (fed cool), or cold gruel of
flour and starch, or eggs — or whisky!

Laxatives include medicinal liquid paraffin for preference, given as a drench, or linseed, castor oil, molasses, hay tea, yeast or oatmeal gruel.

Parasites

The main internal parasites are 'worms' (larval stages) of various kinds and there are eighteen species known to affect cattle. They attack the digestive tract or the lungs, causing a wide range of symptoms but typically general loss of condition, lack of growth, poor appetite, perhaps diarrhoea, and in the case of lung-worms a bronchitic, husky cough.

The effects can be treated conventionally — there is a very wide choice of pharmaceutical anthelmintics, given orally or by injection, and many farmers automatically 'worm' their livestock at frequent intervals just in case, which is quite unnecessary and could indeed be harmful not only to the livestock but also to those who consume their produce. Use garlic to build up resistance.

DISEASES AND DISORDERS

Even in the most carefully balanced and holistic organic system disease is possible. Here are some of the more likely problems and diseases you might be unlucky enough to encounter, though most can be avoided by practising good husbandry. Do not rely on yourself for diagnosis: get an expert opinion, and then decide on the appropriate treatment of the animal and how to eliminate the cause.

ABORTION: Could be brucellosis, leptospirosis, deformed foetus, etc. Isolate until cause is diagnosed (could be infectious, to humans, too).

ACETONAEMIA (KETOSIS): Metabolic disease, mostly high-yielding cows, especially Channel Islands breeds and especially in late pregnancy or at calving; exacerbated if fed too much protein, cereals, kale or silage, or on deficient pastures/hay. Leads to glucose deficiency and accumulation of ketones in tissues (excreted in milk, urine and breath). Symptoms: breath/milk smell of pear-drops, noticeable loss of appetite, reduced milk yield. Prevent with good husbandry — especially if winter-housed (let her out to choose her own green remedy, e.g. ivy) — or give drench of molasses in warm water or inject with vitamin B.

AFTERBIRTH RETENTION: Afterbirth normally expelled within 2-3 days of calving (and often eaten by cow). If hanging out (thin and stringy) for more than four days, get expert to 'unbutton' it carefully — do *not* yank at it yourself (she'll become barren). Everted uterus would be huge, red and heavy: get vet *urgently*.

BLOAT: Accumulation of gases in rumen, which bulges from left flank like a tight balloon. Usually from eating too much leafy and highly digestible pasture, especially with a high clover content, typically in spring when cattle are first turned out on pasture after wintering indoors or in yards. The animal cannot belch to relieve the pressure; the rumen becomes drum-tight and the animal suffers increasing distress, rapid breathing, possibly a protruding tongue and slobbering. Death sometimes occurs within three hours of the onset of the symptoms if no treatment is given. Act very quickly to reduce pressure: encourage the cow to walk about but don't agitate her; give special anti-bloat drench or turps with raw linseed oil, or vegetable oil in warm water, or margarine. Experts can use a stomach tube (but less satisfactory) or, real experts, a trochar and cannula to punch a hole in the animal's side so that the gas can escape — extreme emergencies only.

Better to prevent in the first place: feed fibre (hay or straw) before turning out onto young pasture, or keep animals off clover-rich grazing until later in season and do not let them graze red clover in spring. Limit the amount of time on spring grazing for the first week or two anyway.

BSE: Bovine spongiform encephalopathy, colloquially known as mad cow disease. So far, in Britain, BSE has been almost entirely confined to cows in or from commercial dairy herds.

Because of its sudden spread, far too little has yet been proven about the disease but it is thought that BSE is linked with scrapie, a progressive and fatal degenerative disorder of the central nervous system in sheep which has been known to sheep farmers near the English/Scottish borders for many years. Scrapie can probably be contracted by lambs from their mothers, as they progress along the birth canal.

The current theory is that cattle contracted the disease by eating concentrates containing animal protein which included the rendered remains (meat and bonemeal) from scrapie-infected sheep. The processing of such remains in Scotland was by a process which perhaps dealt more effectively with the infectious element and hence, it is suggested, the very low incidence of BSE in Scotland. The inclusion of such animal protein in concentrates fed to ruminants was banned in July 1988, in an attempt to halt the spread of BSE but the disease was already dormant in many herds.

The symptoms of BSE tend to creep up on you gradually and might be difficult for amateurs to pinpoint. They are largely behavioural and include increasing evidence of *uncharacteristic* nervousness — perhaps unreasonable distrust of coming indoors, a tendency to spook and kick out at nothing in particular, excessive head-tossing, exaggerated alarm at minor noises or normally acceptable circumstances. There might also be constant rubbing without obvious cause, or much tooth-grinding and tongue-chewing, or a general loss of condition and weight. Above all, perhaps, especially in the later stages, there is that pathetic slithering, stumbling and obvious distress you must have seen time and time again on television newsreels whenever BSE is mentioned.

BSE is a notifiable disease (it must be reported immediately to the Ministry of Agriculture and your local council's environmental health officer) and there is at present no treatment, no vaccination to prevent it, no absolute proof that it cannot be passed to offspring and no test to detect it in a living animal. The only confirmation of the disease is from testing the dead animal's brain.

Farmers did not know that infected animal protein was in manufactured concentrates before 1988 and it is still not known how widespread the dormant disease might be. Bearing in mind that ruminants are not carnivores, the safest course for a prospective cow buyer is to purchase only from, say, suckler herds where there is little or no tradition of feeding concentrates (only dairy farmers push their cows to such productive extremes) or from reliable, long-term organic farmers, or from smallholders who, in principle or through lack of finance, have never relied on manufactured concentrates which might, unknown to them, have contained infected animal protein.

BST: Not a disease but is included here because its initials are often confused with BSE! BST is bovine somatotrophin, a naturally produced protein hormone which stimulates the growth of all body tissues in cattle, regulating muscle formation in young animals and influencing the development of the mammary gland. It has been known for half a century that BST can increase a cow's milk yield and in the last few years, to exploit that property, commercial BST has been artificially manufactured by using genetic engineering on bacteria. It is now being used in selected trial dairy farms on an experimental basis to boost yields: it is injected at regular intervals, often fortnightly, and is capable of generating a 20 per cent increase in a cow's milk production. It also increases the fat content of the milk and can apparently raise the level of the hormone IGF (insulin-like growth factor). The injected use of BST can have several undesirable side effects on the cow and, though it is not known if BST-boosted milk affects the humans who drink it, the welfare of cows induced to produce such high yields is surely at stake.

CONSTIPATION: Could be disease, or impacted rumen (give exercise, lots of drinking water and massage). Need to find cause. Can be relieved with Epsom salts, ginger, linseed oil, bran-and-molasses mash, liquid paraffin (medicinal, the kind you use for yourself), pulped carrots, oatmeal, drench of black treacle in water, or put golden syrup in a calf's ration.

DIARRHOEA (SCOURING): Causes might be diet or infection — find out which. Could be sudden change of food, frosted food, too much of something, slight poisoning, allergy to a food. Calf on too much milk usually produces yellowish scouring; deprive of milk for 24 hours and give two feeds of egg-and-glucose in water (4-5 tbspns glucose, one beaten egg, two litres of warm water), then gradually put back on milk (controlled suckling 3 minutes at a time, 3 times a day for 3 days). 'White scour' in calves more dangerous and usually infectious, needs immediate veterinary attention or could die (calf all humped up and miserable). Many other infectious scours, especially in artificially reared calves. Might need oral rehydration therapy. Kaolin helps control scours.

HUSK: Bronchial infestation (calves) by nematode worms: husky cough, breathing difficulties. Prevent by husbandry; can vaccinate if there is known to be a problem on the holding.

MASTITIS: Bacterial infection giving udder inflammation: hot, swollen and tender; lack of appetite and interest, reduced milk yield, characteristic clots in first squirts of milk. Can be highly infectious. Causes various: genetic (some breeds or families more susceptible), dietary (overuse of concentrates, lack of balance), stress of various kinds, lack of hygiene exacerbating a small injury etc. Prevention by good husbandry. If known to be susceptible, give well balanced diet (green food, roots, hay, straw, silage — all home-grown if possible —

and weekly doses of garlic and daily dessertspoonful of seaweed meal). Consult vet in early stages but avoid automatic use of antibiotics to cure. Try this instead:
– Milk out affected quarter at least four times a day
– ease inflammation with alternate hot/cold fomentations
– spray with cold water under pressure (hose) for 10 minutes
– give plenty of garlic (two whole plants a day)

Less drastic: day and night, every few hours (2-3 hours if severe), spray with cold water and strip out milk, dry and massage until cured. If severe, stop feeding for 24 hours but give drench of molasses in warm water.

MILK FEVER (HYPOCALCAEMIA): Metabolic disease, especially in first 48 hours after calving and in high-yielding breeds (Channel Islands often susceptible). Symptoms include paddling of hind legs (slight paralysis), agitation, lack of co-ordination, cold ears (subnormal temperature); can soon lead to going down, lying with neck kinked characteristically; soon followed by unconsciousness and death. Fast action essential: vet will give injection to boost blood calcium instantly. Most common in autumn on fading grazing: give crushed nettles, vitamin D, extra magnesium but *low* calcium diet to prevent, and only partially milk a newly calved cow (milk drains out her calcium).

PNEUMONIA (CALVES): Viral, mild or severe, and many different viruses possible, making calf susceptible to all sorts of bacteria. Avoid with good husbandry and hygiene (problem is mainly with artificially reared calves indoors). Symptoms include rapid breathing, gasping with neck stretched, dry cough, fever, perhaps discharges from eyes or nostrils, tooth-grinding in pain. Isolate and call vet immediately.

RINGWORM: Fungal infection and highly contagious (to humans too). Grey, itchy, circular scabby patches, hair loss. Rub with raw lemon juice, mustard paste or tincture of iodine and seek advice.

SCOUR: See diarrhoea.

STAGGERS (HYPOMAGNESAEMIA): Metabolic disorder with low levels of magnesium in bloodstream, usually at spring turnout if wintered indoors. Nervousness, over-alertness to noises, twitching, trembling, shivering, eyes rolling, staggering; can lead to spasms. Avoid any distress — keep quiet and calm and soothed. Quick action essential to avoid death: vet will give injection to boost magnesium instantly. As a sensible precaution, prevent by ensuring adequate magnesium in the diet *before* turnout.

6. Yield it!

'We have a use for the cow,' said Mahatma Gandhi — and indeed the cow has been exploited for its usefulness ever since it was first domesticated some eight thousand years ago: there would have been no point in domestication otherwise. However, he continued: 'That is why it is religiously incumbent upon us to protect it.'

You might or might not agree with Gandhi's religious beliefs but the essence of his message is that of this book: respect and care for your cow. There are large tribes in Africa who traditionally do this to such a high degree that they hardly *use* their cattle at all — they keep them because they *like* them and they value them as a symbol of wealth, currency rather than commodity, but of such great value that they form the currency of trust.

The main produce demanded of cattle includes milk, meat, manure and muscle-power, though in some places cattle are deliberately bred for the pattern of their hides (as ceremonial shields) or for their horns, or to take part in sports such as bull-fighting and cow-fighting. It is only in the case of meat and hide that the death of the animal is a pre-requisite and, if that should be unacceptable to you, there is ample scope with milk, manure and work to help a cow be self-supporting or even, if you wish, profitable. However, bear in mind that in order to produce milk a cow must first produce a calf and if you do not sell or slaughter the calf at some stage you will soon need extra land to support your expanding herd. If you are very lucky, your cows will only give birth to heifer calves . . .

MILK

We have all, as mammals, depended utterly on milk early in life, whether mother's milk or a substitute reconstituted largely from cow's milk. For a very young stomach, whatever the mammal species, milk is nutrition at its most acceptable and digestible. Adults are less able to digest milk than the young, be they human or any other species of mammal.

Milk is a food, not a drink. It contains:

Proteins (albumin, globulin, casein)

Fats (saturated and unsaturated fatty acids)

Carbohydrate ('milk sugar' or lactose)

Vitamins (A, B, C, D, E and K)

Minerals (calcium, chlorine, magnesium, phosphorus, potassium, sodium and trace elements)

The average proportions of the components in milk are:

Vitamins and minerals	0.8%
Proteins	3.1%
Fats	3.8%
Carbohydrate	4.6%

Most of the rest is **water.**

It is only relatively recently that Europeans have made use of cow's milk for human consumption. In medieval Britain the most common milking animal was the sheep, and here and there people milked goats. There is a considerable difference between the milks of different mammals in their component proportions and in their digestibility.

There is another factor. It is likely that far more people have an allergy of some kind, recognized or not, to cow's milk than to that of goats or sheep. Some people drink goat's milk because

Species	% Fat	% Protein	% Carbohydrate
Cow (Friesian)	3.1	3.1	4.8
Deer (roe)	12.0	7.0	3.6
Donkey	0.6	1.4	6.1
Goat	4.5	3.3	4.5
Horse	1.6	2.2	6.4
Human	2.1	3.8	6.3
Pig	6.3	4.8	3.4
Sheep	7.1	4.2	4.9

they are allergic to cow's milk; some people drink it because they prefer it, but the great majority are more familiar with doorstep cow's milk and, thanks to the very high standards of the dairy industry today, they *trust* it more. Goat's milk has in the past suffered from a reputation for poor taste and smell which is usually more to do with the sloppy hygiene of the goat-keeper than any intrinsic quality in the goat, though some goats will always produce 'strong' milk. As a very young child living in Greece, I was put off goat's milk for life!

Cow's milk, too, can become tainted by careless management and its quality varies anyway between different breeds and individuals within a breed.

The **quality** and **quantity** of milk are affected by:

- Genetics
- General health
- Diet and water intake
- Environment and season of year
- The cow's cycle (length of time since calving)

Some breeds produce much richer milk than others, that is to say the proportion of water in the milk is lower; others (and the relationship between quality and quantity is usually inverse) produce higher total yields, but quantity can be deceptive: the valuable portion of the milk is its solids and usually the higher-producing breeds have a lower proportion of solids in their milk so that you need to drink more of it to absorb the same weight of solids as from a smaller intake of a richer milk from a different breed.

Milk readily picks up **taints** from its surroundings, from the air as well as from direct contact with strong-smelling substances such as creosote, disinfectant, paraffin, fresh paint, kitchen onions. Milk can also be 'flavoured' by what the cow eats, including certain wild plants, some of which also affect butter-making or cheese-making properties. Avoid feeding common milk-tainters (too much kale or turnips, for example) until *after* milking. Certain bacterial diseases affect the taste of the milk, too, making it sour, fruity, fishy, malty, unpleasantly sweet (acetonaemia) or salty (mastitis). If it smells wrong or tastes wrong, it could be an early warning of illness. **Antibiotics** contaminate milk. Injected into the bloodstream, their residues can reach high levels in the milk for several milking sessions after the last treatment. Do *not* sell milk from a treated cow; do not drink it yourself, especially if you are allergic to antibiotics; do not feed it to other livestock; do not try using it to make cheese or yoghurt (antibiotics inhibit the growth of micro-organisms needed in these processes).

Milk is made from cells supplied by the cow's bloodstream to clusters of *alveoli* within the udder which synthesize and secrete the milk at a constant rate for twelve hours after milking; the production rate gradually lessens thereafter. The fresh milk is channelled by a network of ducts into a gland cistern above the teat, ready to fill the teat cistern when stimulated to do so ('let-down') so that the calf can suck it out through the teat's opening.

Let-down is a process by which the milk stored in the udder is released from the gland cistern so that it is made available for withdrawal from the teat. Controlled by hormones, it is a conditioned reflex to a calf's demands for milk. The **stimulation of let-down** is crucial to milking. A milker must evoke a similar response by using a familiar stimulus which the cow associates with being milked. This might be simply the regular routine that includes a familiar milking area at a set time of day, or might need the additional stimulus of udder massage (by washing) or the expectation of food in the milking area, or the presence of the calf.

Milking *must* be a matter of routine for best results — in the same place, at the same time, following the same preliminaries day after day without fail. House-cows not suckling calves need to be milked *twice a day* for the first few weeks or months after calving, ideally at 12-hourly intervals but more practically first thing in the morning and 8-12 hours later in the afternoon. And they must be milked *every single day* during the lactation, which will probably last for at least nine months — make quite sure that somebody (or some calf) is always available to extract the milk for what will probably be some three hundred consecutive days without break. The udder *must* be relieved of its burden regularly. *Regularity* is the key to good milking practice. Set a time-table and stick to it each day. Use the same area, set aside for milking only, whether in the field or in a building.

The less frequently a cow is milked, the less she produces: her body assumes that the 'calf' is beginning to wean itself. If, therefore, you reduce the number of milkings to once a day, she will quite quickly reduce her milk supply and might dry it off completely.

The normal lactation follows a definite pattern in the daily amount of milk produced, whether for the calf or for the pail. From zero just before calving, it rapidly climbs to a peak daily yield when the calf is about three or four weeks old; then it gradually declines as the calf takes an increasing proportion of solid food. In a healthy cow, the lactation 'curve' plotted on a graph is fairly smooth, as the diagram shows. A sharp drop or kink at any stage might be a warning of illness; it might be simply a sign of bulling or reflect some kind of disruption to normal life — for example, being deprived of adequate grazing time, or suffering a degree of stress from being handled by strangers or in the company of unfamilar animals or, of course, being deprived of a calf.

If she does not do so herself, you should deliberately dry off a cow's milk supply a month or two before she calves again, to give the udder a rest and let her build up body reserves in preparation for the next lactation. (That means, incidentally, that if you want a constant supply of milk you need at least two cows and must juggle their calving dates to ensure that at least one of them is always producing milk.)

To dry off the milk supply:

- Reduce to once-a-day milking for a week or two.
- Reduce (or omit) any extra feeding which was providing the production ration.
- Reduce to milking alternate days for a week if the supply is still unmanageable.
- Then *stop milking completely*. She will reabsorb subsequent production of what is by then a low daily yield.

How to milk

There are four teats (or should be!), each with its own milk-factory compartment within the udder so that it is possible to 'milk out' one teat completely without drawing milk from any of the other 'quarters' of the udder. Each quarter is separated from its neighbours by membranes, and the strongest partition is between left and right. About 60 per cent of the milk is produced in the back two quarters.

Milk with both hands, a teat in each, squeezing alternately so that you are milking two at a time. It does not matter how you pair them but most people milk the front two together and the back two together, partly because the latter have more milk in them. There is also sometimes a difference in size between front and back teats.

Once let-down has been stimulated, it is important to complete the milking session while it lasts — ideally within ten minutes, preferably less. But *relax!* If you try to hurry, or if you are totally incompetent or the cow doesn't like you anyway, she will hold up the supply and you cannot — and should not — do a thing about it except give up.

In the 19th century, Bantu milkers used a special cow language as part of the milking routine to encourage let-down. It was a vigorous mixture of shouts, screams, loud whistles and tender words of praise. Europeans were quite unable to milk the native cows because they couldn't speak the 'language'.

If let-down is successful, the milk will come out easily and quite forcefully under your encouragement after the initial few trickles. Don't stop for anything while it is still flowing. It will gradually reduce to a trickle again when the teats are nearly empty and the bag has become soft instead of tight with milk.

It does not, however, pour out of the teat of its own accord but needs to be drawn out, in nature by a calf's sucking and in milking by the pressure of the milker's 'grip', whether manual or machine. 'Grip' is a bad word: it's much too strong for what is involved and would soon damage the udder tissues if taken literally. It is a case of encouragement, not force.

The milker's grip
- Take a teat as a handful, wrapping your fingers around it.
- Apply careful pressure at the top of the handful, using forefinger and thumb to 'trap' milk already swelling the teat.

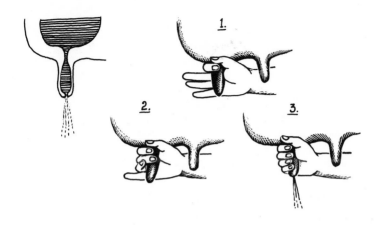

- Maintain this hold while successively squeezing (gently but firmly) with second finger and third, urging the milk down the teat to squirt out into the pail. Relax all the pressure momentarily to let the teat refill, then repeat.

Sit close to the cow on a low three-legged stool (it's easy to set on uneven surfaces, and easy to wheel out of the way quickly) or squat if your legs are used to it. Press your head against her flank so that she knows where you are and you know what she is about to do. Keep your hair well out of the way for the sake of not tickling the cow and inviting a fly-swatting tail or kick. Cows don't trust hesitant movements, be they the tickle of your hair or the too-delicate approach of your fingers to the teat. They like to be aware of what is coming, so be firm but gentle . . . . . .

Some cows are awkward at milking, either from bad experiences in the past, or sore teats, or personal dislike of the milker or sheer bad temper (rare). In some cultures such a cow would be prevented from kicking the bucket by having one hindfoot roped up off the ground — which makes kicking impossible.

When the flow has subsided, gently **strip out** most of the residual milk from the teats. Don't be greedy for every last drop but don't make a habit of leaving too much behind. Strip by sliding finger and thumb gently down each teat, without yanking them down like bits of elastic. Finally, wipe the teats clean and dry. On a dairy farm it is the custom to dip each teat in a special disinfectant after milking to prevent possible mastitis but that should not be necessary if your hygiene standards are excellent and your production demands not excessive. Some people believe that regular teat-dipping *encourages* mastitis by wrecking the teat's natural defences.

Milk is a living substance and can quickly deteriorate when it leaves the udder unless everything with which it comes into contact is as clean as you can possibly make it. **Hygiene** is essential and you should make sure that everything is clean before starting to milk. That includes the floor (concrete is scrubbable), the milk pail (stainless steel is the most hygienic — it can be scoured with boiling water — but rather clangy), your hands, the cow herself and in particular her udder. Wash the udder in warm water, using a clean (boiled) or new cloth every time. With so much close contact, you will automatically notice whether there are any injuries or sore places in need of attention: hand-milking has many advantages over machine-milking. Small machines are available but are not worth it unless you have several cows and are prepared to keep the machine as clean as your hands and milk pail, though they do have the advantages that the milk goes straight into the container (by-passing atmospheric bugs) and the machine can be used by stand-ins who are not practised hand-milkers. Hand-milking, incidentally, like any physical exercise, will probably make your muscles ache until you have been practising regularly.

Milking routine
 1. Clean everything in sight.
 2. Stimulate let-down by familiar routines.
 3. Wash the udder for hygiene and to encourage let-down.
 4. Start to milk within a minute of let-down.
 5. Get close to the cow while you milk her.
 6. Check the first few squirts of milk for signs of mastitis (flecks).
 7. Milk two teats at a time in alternate hands.
 8. Once you have started, finish the job in one go and as quickly as possible.
 9. Stop when the flow stops.
10. Leave the teats clean and dry.
11. Remove, cover and quickly cool the milk.

Milk for sale needs to be even cleaner than house-milk —
definitely no little bits or udder hair floating about in it. Filter it
if necessary. It must also be reduced in temperature as quickly as
possible to less than 5°C and thereafter kept under refrigeration.
Don't freeze it, though: unlike sheep and goat's milk, cow's milk
will separate into unpleasant flecky bits and globs when it thaws.

If you do want to sell the milk or dairy products, you need to
register your intent and will be subject to inspections of the
premises and of the cows by environmental health officers and
Min. of Ag. vets. To protect the public, you'll find yourself
wrapped up in plenty of red tape and it is hardly worth the hassle
on a very small scale. Consult your local Milk Marketing Board
or Min. of Ag. and local council for full details. You might need
to heat-treat the milk to improve its keeping qualities.

Raw milk is untreated, fresh milk with a maximum of vitamins,
flavour and natural enzymes.

Pasteurized milk has been heated to 63°-66°C for half an hour to
kill pathogenic organisms.

Bearing in mind that even an unpressured house-cow produces
many pints of milk every day, you are bound to have a
substantial surplus unless she is also feeding at least one calf.
Lucky you: convert it into substances which will keep much
longer than raw milk does and which also leave you with
quantities of diluted milk (buttermilk, skimmed milk and whey)
which you can use as a very good feed for the pigs and poultry
you suddenly find are a good idea after all. Cows are so generous
that they inspire you with ambitions to expand . . .

The conversion of dairy produce is a subject as complex as
making wine and the Bibliography suggests some comprehensive
books on the ancient art and craft of the dairy. There is space
here for only the briefest details of the basic processes to give you
some idea of what is involved.

How much milk? Depending on quality, you need about:
8-9 litres (14-15 pints) (Jersey) for 1kg (2lb) clotted cream
14 litres (25 pints) (Friesian) for 1kg (2lb) double cream
20 litres (35 pints) for 1kg (2 lb) butter
10 litres (17 pints) for 1kg (2 lb) hard cheese

Cream

If you let fresh milk stand for a few hours in a wide vessel, the cream rises naturally to the surface and the longer you leave it the more cream you can 'set' so that it can be skimmed off. Cow's milk is unusual in this respect; goat's milk, for example, is homogenous and the cream does not rise and separate easily because its fat globules are small. Fat globules are lighter than the rest of the milk and rise by about 0.6mm an hour — the bigger they are, the faster they rise. Cream contains most of the milk's components but in different proportions to whole milk. In particular it has a higher fat content.

To squeeze the greatest amount of cream from the milk you can use a mechanized separator, which works on the principle of centrifugal force, flinging the fat globules together 6,500 times faster than natural coagulation by rising.

- **Double cream** contains at least 48 per cent milk-fat.
- **Single cream** contains at least 18 per cent milk-fat.
- **Clotted cream**, containing at least 55 per cent milk-fat, is set for up to 24 hours in a large, shallow pan which is then carefully transferred (without skimming) to be very slowly heated to 82°-85°C and maintained at that temperature for about 45 minutes so that the cream surface wrinkles and can be skimmed off after cooling overnight.
- **Soured cream** is either untreated cream left out of the fridge for a few days or can be made instantly by adding a little lemon juice or commercially by adding a starter culture to pasteurized cream to convert lactose into lactic acid.

Ice-cream

Experiment by following a good cookbook. Commercially, mix raw milk with cream, sugar and flavourings (and eggs if you like), heat them to pasteurize to the food inspector's standards and then freeze the mixture, while occasionally agitating it.

Butter

Set and skim the cream at room temperature and find another use for the skimmed milk. The main steps in converting cream into butter are:

1. Agitate the cream until the fat globules come together to form yellow granules.

2. Drain off the buttermilk and thoroughly rinse out any residual milk.
3. Work the butter to squeeze out every last drop of moisture, using palette knife or wooden 'Scotch hands' paddles. Add a little salt if you wish.

The main problems arise in getting the granules to 'come' during the churning and in getting rid of excess moisture — wet butter will not keep at all. Churning sometimes takes as long as half an hour — hard labour if you are hand-churning but it gives you time to meditate until suddenly, just as you were about to give up, the butter granules begin to form.

Churns
- For fun, simply shake some creamy milk vigorously in a sealed bottle: it will eventually come.
- Use a food mixer for small quantities but with care — it'll be a blade-binding butter-lump before you can blink.
- Use a manually operated churn: a glass Blow churn which works like egg-beaters (hard work but relatively cheap and with a choice of sizes) or a larger and more elaborate wooden churn — there are many different patented designs of greater or lesser beauty and ingenuity.
- Use a churn driven by electricity, a treadmill dog, a miniature windvane, solar energy or water power — invent something.

Yoghurt
Use whole or skimmed milk. Heat it carefully to just below boiling point, cool quickly to hand-heat, mix in a starter culture, incubate in a warm place in sealed containers until the curd has formed, then cool it and refrigerate. **Starters** are bacterial cultures which activate the making of lactic acid from lactose (milk's natural sugar) and turn it sour. Raw milk sours naturally but heat-treatment kills useful as well as unwanted bacteria; the former need replacing with a starter to curdle the milk. Starter cultures for yoghurt can be bought for the purpose (liquid or freeze-dried) or can be from ready-made yoghurt.

Soft cheese
Soft cheeses do not last more than a few days but they are simple to make. For **cream cheese** at home, just ripen some raw cream by leaving it out of the fridge for a while to sour naturally, then

drain it in cheesecloth or butter-muslin. Safer cream cheese is made by heat-treating fresh cream, adding it to milk still warm from the cow, cooling the mixture to hand-heat and adding cheesemaker's rennet so that it forms curds, which can be strained and left to ripen. **Rennet** is an agent used to coagulate curds in making cheeses. It contains rennin and pepsin. **Rennin** is the enzyme within a calf's stomach which clots the calf's milk to make it digestible. Rennin comes from newly-dead calves but you can also buy vegetarian rennet derived from fungi, or use curdling plants like lady's bedstraw.

Cottage cheese is made from milk rather than cream. Warm it to 38°C, add lemon juice or vinegar, leave for 15 minutes to sour and form curds and whey (yes, Miss Muffet). Strain through muslin, letting the bag drip for a day or so. Rennet would speed up the whole process.

New ideas
As milk is such a valuable food, it is worth scouring the world for ideas about converting it into interesting cheeses. Look much further afield than your local supermarket; look at other European countries and further still — you'd be amazed at what they do with milk in Asia and Africa. Next holiday?

ORGANIC MEAT

You may not intend to produce meat but, as luck will have it, some bull calves will arrive . . . Apart from broad standards for organic meat production (including welfare, housing and diet), the following points apply specifically to beef animals in organic systems, either in accordance with UKROFS (UK Register of Organic Food Standards) or based on practical experience.

Choose breeds and strains of cattle suitable for local conditions and organic systems. For single-suckler cows you do *not* want high-yielding dairy types but you *do* want naturally good mothers who are interested in looking after their calves — hill breeds are often the best. Let breeding cows remain in a fairly stable group and let suckled calves stay with their mothers until they are weaned naturally by the cow about a month before she calves again.

Keep the herd 'closed' to avoid buying in disease but, even if you have your own bull, use AI from time to time to broaden the genetic base and avoid inbreeding. If you do acquire stock from elsewhere, the animals should have been born and raised on a known organic unit. Calves for finishing (i.e. taken to a suitable weight and condition for slaughter) can be acquired at normal weaning age from suckler herds, or from dairy herds if they are no more than 28 days old, have received colostrum for at least the first four days of life and have subsequently been fed on milk that contained no antibiotics or growth promoters. If the calves are more than two weeks old, they must have had daily access to dry food containing sufficient digestible fibre to aid development of the rumen.

Artificially reared calves should be fed on fresh, whole, organically produced cow's milk, fed through a teat to satisfy their urge to suck, until they are eating solid foods. Only use milk replacers in absolute emergencies. Don't wean until the calves are at least nine weeks old.

As far as possible, systems for suckler herds, store cattle (i.e. beef animals who are merely 'ticking over', usually on cheap winter feeding before being turned out to graze in spring, when the improved diet will give them a compensatory burst in growth rate) or finishing cattle should be based on grazing, though you can also finish beef animals in well bedded, spacious yards.

Keep stocking rates low. Depending on the land's condition, on permanent pasture allow about one hectare (2.5 acres) for each cow and her calf — this should give enough space for making hay or silage as well as for grazing without risking a parasite overload. In winter, give hay and silage but also allow access to grazing whenever field conditions permit, and let the animals graze or seek shelter by *choice* throughout the year.

Let the cows calve down outside unless they are first-calf heifers or you expect any problems. Beware of calving single-suckler cows between late April and late June because they will produce too much milk for one calf and you will need to milk out one or two teats until the calf is big enough to drink more. Don't feed a cow cereals unless she has twins but do feed hay and straw in the field in the autumn as well as winter hay or silage.

Don't be too eager to get *heifers into calf* at an early age — they should be properly developed first (eighteen months at conception at the earliest, preferably two or two and a half years). *Separate yearling heifers* from the herd for a few weeks while the bull runs with the cows (to avoid the heifers being served too young), then let them rejoin the herd and keep the best of them for breeding later on. Then let a first-calf heifer have her second calf 14 months later rather than 12.

The use of feeds *containing non-food ingredients* intended to stimulate growth by modifying the gut microflora or the endocrine system is prohibited. Avoid the *routine* use of *anthelmintics*, and only administer *antibiotics* if and when absolutely necessary. Do not subject the animals to any *surgical or chemical interference* except to improve their own health or wellbeing. Finally, use a *trusted* and *local* abattoir for slaughter and make arrangements to ensure that the animals are slaughtered soon after arrival.

The end of the line

As in life, so in death should concern for the animal be your top priority. The trouble with rearing animals for meat is that they have to be killed. It is a reality you must face and come to terms with but at least you can take every care to ensure that death is quick and kind, and that the stress immediately before death is minimized. Those who scoff at putting welfare first might bear in mind the practical and economic fact that any butcher can recognize meat from an animal which has been killed in a state of distress.

The best situation, from the animal's point of view, is that it should be killed at home by an expert rifleman while it is calmly grazing in a familiar field. It would know absolutely nothing about the whole business at all. It would certainly avoid all the stress of being transported to an alien place where lots of other displaced creatures are waiting to be slaughtered, where the smells and noise and activity are distressingly unfamiliar and where it is quite clear that something out of the ordinary is happening. It is said that animals have no anticipation of death and therefore do not fear it, even in the abattoir, but they can certainly be alarmed by the strangeness of the environment.

If an animal is intended entirely for home consumption, it can legally be killed at home, but if any part of it (including offal) is to be consumed by other than your immediate family, whether paying guests or sold to the general public, the animal must by law be killed at a licensed slaughterhouse and must not be consumed unless it has been stamped by a meat inspector.

There is very good reason for this legal ruling. If you are

willing to take the risk of eating what might be diseased meat from your own animals, that is your own concern, but you must not feed anybody else on it, whether or not you are aware of the presence of disease. However good your husbandry, there are some conditions that can occur regardless. On the whole, however, if the animal appears fit and healthy at the time of being killed, there should not be much wrong with its meat — but be aware of the risk, and learn to recognize disease for your family's sake. At the slaughterhouse, experts will detect signs that were quite undetectable in the living animal. Look particularly at the liver — but first of all take a meat inspection course at a college so that you know what you are supposed to be noticing.

Home-killing

The Humane Slaughter Association (HSA) has been working for several decades to improve conditions in slaughterhouses, with the emphasis on the welfare of the animals, and they also give a great deal of very sound advice on the subject of killing your own animals on the holding. The first point to bear in mind is that *cattle are large animals*. It is not as easy to kill them kindly as smaller and more manageable livestock. They need a greater degree of restraint (which can itself be stressful for them) and they are even more bulky and awkward to handle when dead.

Think very hard about this aspect. Are you physically capable of dealing with a large carcase, let alone competent to kill the animal in the first place? And how are you going to dispose of the waste? There are new EC regulations which could affect you in this respect.

More important is the question of how the animal should be killed, and by whom. The essential qualifications for a person who kills livestock at home are that the person should be experienced, accurate, confident and coldly efficient. This is one of those situations in which gentleness is less kind than ruthlessness. However good a shot you are, an animal will immediately sense the slightest lack of confidence or will, the slightest tremor of sympathy or uncertainty on your part, and it will become suspicious and possibly agitated. Again, however good you are at killing, it can go unexpectedly wrong even for the steadiest and most experienced, and if the animal involved is one that you know well the event is doubly traumatic. A trusted but uninvolved third party will have far fewer qualms about ending the life of a familiar animal; it is much harder to kill a friend than a stranger. *If in any doubt at all, don't do it yourself.* Never be ashamed to pass the buck. The animal's peace of mind is far more important than your own foolish pride.

There is nothing to prevent you from asking a friendly local gamekeeper or forest warden to carry out the deed, especially one who has plenty of experience in the humane killing of deer, but it is probably wiser to talk to your vet in the first instance. A vet might be able to tell you where to undertake appropriate training to do the job yourself, or might carry out the job personally in some cases (for a fee) or, more likely, will be able to recommend a local slaughterer prepared to come out to you and deal not only with the killing but also with handling the carcase.

What, then, about the choice of weapons for humane killing? You would be well advised to discuss this question with experts such as vets, the police, UFAW (Universities Federation for Animal Welfare), HSA and the only gun company that makes suitable weapons for slaughter (Ackles & Shelvoke of Tulford Street Works, Aston, Birmingham B6 4QA — tel. 021359 3277/8).

On no account should a shot-gun be used; for a large animal the choice is between free-bullet rifles and pistols or captive-bolt pistols. The former need even more expert handling than the latter but, whichever is used, you will need a firearms certificate from the local police. The procedure can take literally months and they will need a lot of convincing about why a private individual should be allowed the weapon.

A vet would probably use a **free-bullet pistol** but if an animal is being killed for meat rather than simply being put down, it will have to be bled. That means that the most appropriate weapon is a **captive-bolt pistol**, which uses blank cartridges to launch a retractable captive metal bolt into the animal's brain to stun it. Death is not from the weapon but from loss of blood after the animal has been stunned, which means that its throat must be slit. Be warned that this is a thoroughly unpleasant business and that it is also a messy one: there will be a great deal of blood.

If you intend to do the deed personally, you must be thoroughly trained and familiar with the weapon. Practise extensively before you even contemplate killing a living animal. A captive-bolt pistol is fired with its muzzle in firm contact with the animal's head: practise firing into a dead animal or a thick telephone directory. The point of aim must be precise. Visualize the crossing point of two diagonal lines drawn from the base of each horn on one side to the inner corner of each eye on the other.

Unless the animal is to be killed by an expert shot with a rifle from a distance, it will have to be restrained — preferably in a manner and place with which it is already familiar. Remember, however, that there will be a great deal of blood and that the carcase will be awkward to handle. Think about how you are going to get it out of the place of confinement to a clean area for dressing.

Be prepared for the reality of death and know how to recognize it: no rhythmic breathing, glazed fixed eyes, and the tongue lolling from a relaxed jaw. Remember that the captive-bolt does not *kill*. You must bleed the animal to death as soon as it has been properly stunned. Watch out: there will be certain reactive twitches and dangerous leg-kicks from the unconscious animal which will probably continue during the bleeding.

To bleed: Cut the throat from ear to ear, exposing the spine and severing *all* the major blood vessels. The blood will come out rapidly and in quantity for about two minutes and must be collected for safe disposal. (Note that there are regulations governing the disposal of waste, be it blood, entrails, bones or any other part of the carcase.)

Now you know about the reality of meat.

However distasteful it might be to you, the acquisition of good killing skills will help you to end an animal's life in the kindest possible way with the least stress, and that includes peacefully putting down in a friendly environment any animal (domesticated or wild) who is severely ill or old enough to want to die. You could save it a great deal of distress if you have the courage of your principles, and perhaps you owe it to an animal who has been your friend and generous supplier to give it the coup de grace yourself. Never send an old friend to the knacker.

Slaughterhouses

You might think that doing it yourself is too difficult, and a slaughterhouse would be easier. Now for the problems. The ideal slaughterhouse is small and local — perhaps a family butcher who is licensed and prepared to slaughter one or two animals now and then. The advantages are that the journey is cut to a minimum, you can get to know the slaughterer as an individual, you can observe his methods, talk through the procedures and your concerns, and be assured that your animal will not be pressed into too small a pen with a large number of strangers. On a small scale neither you nor your animal will be just a number in a crowd and you will be able to make careful arrangements to ensure that it is slaughtered very soon after arrival to avoid at least some of the stress of being in alien surroundings.

There is, however, a huge vulture hovering on the horizon, with 1992 stamped on its beak and wearing the familiar plumage of European bureaucracy determined to create uniformity. Proposed regulations are almost guaranteed to close down exactly those friendly, low-key, local butcher-slaughterers who are the smallholder's answer to a prayer. It seems that the EC wants them to install the same facilities as the production-line abattoirs that handle hundreds of animals at a time and have no interest in ones or twos or small batches, nor always take care to give you back your own animal's carcase and offal rather than somebody else's. There are about 850 licensed slaughterhouses of various sizes in Britain at present but it is anticipated that, after 1992, there will be only about 300 — and the vanished ones will include the local man near your holding. The great majority will not be able to afford the overheads nor compete with the big places, and you will have to transport your animals and carcases a long way to and from a central abattoir where the welfare standards are bound to creak under the pressure of high throughput. In the meantime, however, your task is to find an abattoir that meets your standards.

To choose an abattoir:
- First approach your local council and neighbouring councils: they are the authorities who license every abattoir in the country.
- Consult other livestock owners for their personal recommendations.
- Consult your vet.
- Talk to farmers and to the abattoirs individually to find out exactly what they are prepared to handle and how.

Also consult the UKROFS (United Kingdom Register of Organic Food Standards) standards concerning loading, transport, lairage and welfare at slaughter. At some stage talk to the HSA, who have a pretty shrewd idea of the reputations of abattoirs all over the country.

Abattoirs, though all are licensed, vary widely in their efficiency, their capacity, their methods and their charges. Start by looking at those closest to home — within a five-mile radius, then ten miles and so on; proximity is an important factor. Then ask some of the following questions:

- Do they deal with cattle anyway?
- What slaughter methods do they use?
  (Captive-bolt is what you want for cattle)
- What are their charges?

The latter can be an important factor and will probably rock you back on your heels: some of the larger abattoirs find small numbers disruptive, especially if you try to set certain welfare conditions, and they seem to scare you away by price. They will certainly charge more if you actually demand to have your own animal's carcase back. They will automatically keep the hide, too, which is worth £25 or £30 on the open market. In the old days, the hide covered the slaughterhouse's costs but today they often keep the hide and the offal, and on top of that charge you for the carcase's hanging space.

When you have narrowed the choice, make a point of visiting the abattoir while cattle are actually being slaughtered. Sometimes such a visit arouses immediate hostile suspicions from those who dislike the activities of animal rights campaigners but, on the other hand, if you are a bona fide potential customer you have every right to distrust an abattoir that will not let you watch slaughtering in progress or which refuses to let you watch your own animal being killed. Such an abattoir is unlikely to be co-operative in every respect.

However much you dislike the idea, you owe it to your animals to assure yourself that they will be treated as well as possible in the circumstances. In fact, any meat-eater should visit an abattoir to see how meat is produced, and I promise you will never forget the sight of heaped horns, piled-up hooves and the mournful eyes of decapitated cows' heads. You'll remember the smell, too.

Steel yourself to watch for at least half an hour and don't make an immediate judgement: some slaughterers are put off by observation and are anyway expecting criticism. But trust your gut-feeling about the place. Look for older, experienced handlers with a quiet manner who don't need to flaunt their manhood by pretending loudly to be callous. The HSA suggests that you check on the following points during your visit:

- Is the **unloading** area secure so that animals cannot escape? How are the animals handled from the vehicles into the lairage? Are stockmen noisy and rough, or quiet and calm? Do they use sticks and goads unnecessarily?

- Is the **lairage** atmosphere quiet and calm so that the animals can settle? Are **handling facilities** well constructed and maintained, with no sharp corners or projections to cause injury? Is the floor likely to cause an animal to fall? Is the building well ventilated, and lit well enough to be able to check the animals for injury or sickness?

- Is the **drinking water** clean and accessible? Are **hay** and fodder provided for overnight lairages? How well or badly or indifferently are the animals being treated in the lairage?

- Is there a **stunning pen**? (There should be for cattle.) How noisy is the pen and how easily are the cattle moved? Does the slaughterer let the animal settle in the pen? Is the **gun** well maintained and is the correct charge used? Does the gun work every time and is there another gun to hand for emergencies?

- How accurate is the **shooting**? Look along the head rail: the bolt hole should be at that crossing point described earlier. Is the animal properly **stunned**? How soon after being stunned is it **bled**? (It should be without delay: death is caused by blood loss and it takes perhaps two minutes for the animal to bleed out.)

- Would you be happy to send your own animals to this abattoir?

A **carcase** must be chilled for at least 24 hours in the slaughterhouse before it can be removed. You will probably be charged for the hanging space, and also for basic butchery to reduce the carcase to manageable hunks.

A butcher-slaughterer who buys live cattle from market for his own shop would always rest the animals for a week to 'let the muscles relax' before killing. If he is eating the meat himself he'll probably buy heifers rather than steers; steers are leaner (as current fashion dictates) but the fat on a heifer decidedly improves the taste and juiciness of the meat. He would also be selective about the breed, probably preferring Angus or Hereford, or crosses from bulls of those breeds. After slaughter he would hang the meat in the chiller for at least ten days, preferably two weeks, to bring out the flavour. Finally he would do all in his power to ensure that his expertise and efforts, and those of the farmers who bred and reared the animals, are not wasted by the cook. Good meat deserves proper cooking!

THE WORKING COW

Cattle are used primarily for work all over the world. Milk is often incidental, and traditionally the animals are only eaten when they have become too old or unfit to work — when the meat is no doubt as tough as the old boots made from their hides. In many cultures, it would seem a huge waste of animal energy to kill and eat a young animal who could have many more productive years ahead of it as a worker. Once it's dead, that's it: the meat might give humans an intake of energy so that they can work, but isn't it easier to keep the animal alive and let *it* take the burden?

The typical working animal (in Britain as well until later in the 18th century) is the ox — a castrated male, who grows much bigger and is far less temperamental than a whole bull. Oxen are massive, ponderous, slow but steady workers trudging the fields and roads of Africa and Asia in their millions in the service of humans. They are bred with work in mind: they are veritable powerhouses at the front end, with good strong necks, chests and shoulders, but usually rather skimpy in the hindquarters (which is where the best meat comes from) because all the strength is needed at the front, which takes the strain of the yoke or the shaft-harness. Their feet are large and sound; their temperaments are tractable and immensely patient. But in many places massiveness is not essential; indeed in hill country it is a drawback and here the typical working animal is a small, even dwarf, cow who supplies her owner with milk and offspring as well as musclepower. In Nepal, for example, they have crossed Jersey bulls on the tiny local black hill cattle as the perfect compromise.

Peter Reynolds, that affable, bearded classics graduate and ethnographist who combined his fascination for research into the prehistoric past with the practicalities of agriculture by running the Butser Ancient Farm research project in Hampshire, has for many years studied the Iron Age in particular and used a combination of scholarship and imagination to re-create the period's farming practices. He has concluded that cattle, or more specifically cows, provided the main traction power for Iron Age farmers. By Reynolds's reckoning, cows might not be as powerful as oxen but they do give milk as a bonus, which means that your own affable house-cow could be persuaded to do a little work around the place and still provide you and her calf with sustenance. In 18th-century England it was almost as common to work spayed cows as castrated oxen (read Arthur Young for details on training and management at the time).

In Britain there are people who are seriously reverting to working cattle. For example, there are several teams of big red Sussex oxen working in southern counties and an intriguing team of two Channel Islands bullocks and a Welsh Black/Kerry cross worked by James Benedict Bygott-Webb in Somerset. He has even established a British Society for the Preservation and Promotion of Working Oxen at his farm near Churchinford. At Butser, Peter Reynolds works Dexter cows, which he thinks are pretty close in size and type to Iron Age cattle, though he uses the longer-legged Dexter rather than the genetic dwarf. Horns are invaluable in working cattle in that they are good anchorage for a steering rope and also prevent a yoke or collar from sliding off over the animal's head.

Cattle can have minds of their own and they need careful, quiet training to work — especially cows, who are usually more intelligent, more alert and more questioning than oxen. Training needs considerable patience and plenty of unhurried time, starting when the animal is very young indeed by getting it used to being handled and led on a halter.

Training *must* be based on trust and encouragement, never on fear; there is absolutely no excuse, ever, for hitting an animal and indeed such violence, however well controlled, is more likely to make the trainee stubborn than willing. If punished, the animal soon learns to view training as something unpleasant to be avoided. Training can only succeed if the animal is willing to co-operate and finds the whole procedure interesting, even enjoyable, while it becomes increasingly familiar. The most important tool in training an animal is your *voice*.

Another positive factor in training is learning by example. A young animal can learn a great deal by watching another of its own species or being teamed with an experienced older animal, and this principle also has the great advantage that the animal remains part of a herd. Farmer and horse-breeder Mr C. 'Trust, not trauma' Pinney, who organizes courses on handling heavy horses, practises a philosophy which is equally applicable to the training of working cattle, along the following principles.

- **Catch them young.** 'In an ideal situation, the easiest time to train a cart-horse is when it is born,' remarks Pinney. He works on the basis of gentle, sensible handling right from the start, using the young animal's natural curiosity about life in general to make it hand-tame, beginning to handle it as early as possible without upsetting the bond between mother and offspring.
- **Let the mother set the example.** Pinney's first training step after the early handling is to let the young animal become used to a halter in its mother's presence, tying it by the halter in the stall with its mother on a lead long enough to let it suckle easily but short enough to avoid entanglement. He grooms the mother as usual so that her little one sees grooming as an everyday and enjoyable procedure; gradually he is able to groom the youngster as well, while it is tied up, and in due course he can pick up its feet, check its mouth and so on without causing any concern. Because its mother is always there, it does not associate the halter with fear or prolonged separation from the parent.
- **Remember it is a herd animal.** A young herd animal feels safe within the herd, even if the rest of the herd is no more than its own mother. It gradually learns about herd life by association with other animals, absorbing its features which, in the ideal training situation, includes the everyday business of its mother's work. And because the mother accepts your presence so readily, the youngster accepts you as part of the herd, another feature of everyday life. As long as the mother shows no fear or distrust of you, so too the young animal will not be afraid or wary of you, and you have established that essential trust which forms the very basis of a good relationship.

Second best to mother as a role-model is an already trained and experienced older member of the youngster's own species. Your biggest problem, of course, is your first trainee, with no example to follow, and that simply means that you yourself are its sole guide. The major principles in this case (and indeed when the working mother is around) are trust based on familiarity and good associations. The animal needs to accept you as a fellow in its herd — a senior fellow, but a familiar rather than an occasional visitor. Spend a great deal of time with the animal simply for the sake of being present, not actually doing anything in particular but becoming a part of the scenery. It is important that you become a friend rather than a master, but a friend higher up the herd's social hierarchy. You want to establish such an empathy that the animal *knows* you will never require of it anything it would find alarming, dangerous or beyond its abilities. Make quite sure that the animal always knows that you are approaching and knows just where you are, especially if it is restricted; use your voice, put a firmly reassuring hand on its rump as you move about behind it, and avoid sudden, unexpected movements.

In this respect cattle are easier to handle than horses: they are much less flighty, much more easygoing about life in general, be they cows or oxen. They are really far too lazy to panic unreasonably and, as long as you have their own type of patience, they will happily plod about their work for ages, deliberate, even ponderous in the case of oxen, but reliable and effective in the long term.

When it comes to asking the young trainee actually to work, continue the kindness and fellowship and guide it mainly with your *voice*. Right from your first meeting with an animal, *talk* to it; let it become thoroughly familiar with your voice and gradually let it understand the implications of different intonations (actual words are irrelevant). If it is ever truly necessary to express your anger, do so with your voice rather than by physical assault. The voice should be your main method of control, not the rope and certainly not a goad, cane or whip. A steering rope should be used only as a gentle reinforcement of the verbal command.

Let every stage be *gradual*, right from the start. Never push an animal beyond its threshold of interest or physical ability. Take the whole business of training, for whatever purpose, at a leisurely pace. Introduce new activities for short periods at first — *little and often* is the key. Be very careful· to introduce draughtwork of any kind gradually: never put too much physical strain on a young animal whose skeleton and muscles are still developing. Moderation in all things. And, lo! you have an ox or a cow content to do whatever you require, be it trundling about the place with a cart, ploughing a furrow, carrying a pack or letting you take the weight off your legs for a plodding ride home.

A tractor, someone once said, is born adult. It is so much more satisfying to know a living creature right from birth and to help shape its role by gentle training as it grows up and becomes your willing ally. You can't talk to a tractor, or scratch its favourite tickle-spot, or cuddle up against it on a cold night. Nor can a tractor turn your excess grass into manure and deliver it to the field while it ploughs.

MUCK

An organic holding could almost justify keeping cattle purely for their manure and ask nothing more of them than their dung to fertilize and restore balance to the soil. It is the satisfaction of fertility by 'completing the circle of growth and returning to the earth organic matter that has served its immediate purpose,' as Laurence Easterbrook put it. Dr Rudolf Steiner was firmly of the belief that only living things can produce life, and manure has that huge advantage over the bagged chemicals that, in theory, provide all the minerals a soil should need.

- One **dairy cow** weighing about 500kg (78 stone) excretes about 41 litres (9 gallons) of slurry (faeces and urine) a day.
- One **beef bullock** weighing about 400kg (62 stone) excretes about 27 litres (6 gallons) of slurry a day.

The composition of cattle manure:

	Dry matter	Nitrogen	Phosphate	Potash
	%	% N	% P_2O_5	% K_2O
F.Y.M.	25	0.6	0.3	0.7
Fresh slurry	10	0.5	0.2	0.5

Proportion of total nutrients available to crops (manure spread in spring):

F.Y.M.	25%	60%	60%
Slurry	30%	50%	90%

Cow dung is basically plants which have been composted by the cow's chewing, body heat and digestive bacteria. Fresh dung, however, is too potent: it can burn off plants and needs to be matured before being used in quantity as a soil enricher.

Farmyard manure (or FYM) is dung and urine absorbed into plant-based bedding, preferably straw. It not only supplies soil nutrients but also helps to adjust the physical condition of the soil in terms of aeration, drainage, water-retention, friableness and the ability to warm up faster in the spring.

To 'make' manure you need to:
- Keep the **heat** in.
- Keep the **rain** out.
- Let the sides **breathe.**
- **Retain** the goodness.

Build a big heap of steaming manure in a stack, either self-supporting or contained by walls with adequate air spaces, and give it a roof — old doors, polytunnels, polythene sheeting, or straw 'thatch' and soil — to keep out the rain and thus avoid dilution. Don't let the goodness leach away from the heap: channel any juices into safe storage as liquid manure, again without letting rain dilute them. Include layers of wetted, bruised, fresh green nettles to boost the heat in the heap. If it stops cooking, turn the heap after two months to get it going again, otherwise leave it undisturbed for 4-6 months. By then it should be a rich, black compost ready to spread among your crops.

To test manure's maturity: Plant some cress seeds in a sample of manure. They will not germinate if it is not ready.

7. And Finally . . .

Do you know any old cows? Seriously, try to think of the oldest animal you have known. About ten years old, perhaps? Twelve, going on thirteen? Hmm. The natural lifespan of a cow is about twenty years.

Why, then, do you not see old cows more often? You might find some out on the hills in suckler herds, or tucked away on smallholdings, but never in dairy herds, where efficiency has to be given top priority.

- 25% of dairy cows are culled before they are forty months old.
- 25% of dairy cows live more than seven years, but not *much* more.
- Most dairy cows have been slaughtered by about five years old.

[Source: CIWF]

The inputs for commercial dairy farming are so high that any cow who slips below high production standards finds herself facing the captive-bolt pistol before she has had time to belch. If her milk yield has fallen below a certain threshold, if she has taken a dislike to the stock bull or the AI operative and failed to conceive first go (second at most), if she has succumbed to the stress of intensive production and developed mastitis in an overworked udder abused by the machines, or cysts on her ovaries, or persistent lameness, or abortion, or — cruelly — if she has suffered the terrible indignity of BSE, then it is the meat-hook for her and no pleading, be she only three or four years old, a mere teenager in a human lifespan.

It does not *have* to be like that. You might be unlucky and lose your cow to an unavoidable disease or, worse, be forced by the authorities to kill her in case she is a carrier; you might lose her to a genetic time-bomb or an unfortunate accident; but with reasonable luck and a great deal of caring respect, diligent stockmanship and sound husbandry, your cow could be with you for far more years than the family dog and will still be there when the children have left home. If conditions are right, she will go on giving you milk until she's twenty or more.

When I think of how few cows die at home in peace, I am appalled at the harshness that executes animals because they are no longer *useful*, because they have become a liability instead of a financial asset. What a reward for a productive life! You, even if you reject all this sentimental twaddle, have more of a choice than someone whose living depends on the herd's balance sheet. You, the realist, can have the satisfaction of knowing that you have given your animals a good life and the chance of living out their span until their health or sheer old age determine that their time has come. You, keeping cows for the pleasure of their company, can afford to indulge them.

An ageing cow is naturally more likely to develop various aches and pains, particularly in the joints, and you need to be alert for her discomfort. Also check the state of her teeth in case she has difficulty chewing harder foods like hay, and make sure that her feet are kept properly trimmed. Give her the option of taking shelter when she wishes, be it from hot sun, rain, wind or cold, and don't ask her to walk too far to take advantage of it. Let her choose her own grazing, exercising her years of experience. Let her decide for herself when she no longer wants to bear and raise a calf or produce milk for you. Let her have familiar company even if she is a bossy old thing. Let her be pampered a little.

When her last moments finally become inevitable, continue that tradition of respect for her dignity and concern for her peace of mind right through to the end. On no account send her away to be slaughtered if killing is the kindest decision. Talk to your vet and decide the best way for her to go — the best for *her*, that is, but, whatever the method, let her die at home, in a place where she has spent so many good years.

And, do you know? You'll miss the old thing when she's gone even if she was cantankerous and stubborn and a bloody-minded matron who always wanted (and got) her own way in all things. You'll miss her even more if she was the typically affable, generous, easygoing, affectionate, self-possessed and humorous character that most cows are at heart.

Bury her deep so that she remains undisturbed. There is a little dell-meadow beyond the walled gardens of a large old manor house which I visit often — the sort of place where you can sprawl undisturbed, wasting time, at peace with yourself. It was several years before a very old lady let me in on its secret. It had been the last resting place for the big house's favourite animals — their dogs, their cats, the children's hamsters and ponies, the faithful cart-horses and several pet Jersey cows. It was a soft, quiet place, gently draped with a great calmness and tranquillity such as you so often see in the faraway eyes of cows. Bless them all!

Cow facts

Normal temperature: 38°-39°C (101.3-103.1°F)

Normal respiration rate: 12-15 per minute

Normal heartbeat: 45-60 per minute

Normal teeth (adults): — 8 incisors, 12 premolars, 12 molars
(no incisors on upper jaw)

Possible age of heifer at first bulling: 6-14 months

Minimum age for first service: at ⅔rds of mature weight,
probably 15 months old — but much better to wait until she is
well grown, say 2 years old

Duration of oestrus: 6-26 hours

Time of ovulation: 10-15 hours after end of oestrus

Interval between heat periods: normally 18-24 days (typically 21
days)

Average gestation: 280 days

Useful addresses

British Association of Veterinary Homoeopathy
  Chinham House, Stanford-in-the-Vale, Faringdon, Oxon,
  SN7 8NQ

British Organic Farmers: see *Soil Association*

Broad Leys Publishing Co. (*Home Farm* bimonthly magazine)
  Buriton House, Station Road, Newport, Saffron Walden,
  Essex CB11 3PL

Compassion in World Farming (CIWF)
  20 Lavant Street, Petersfield, Hants GU32 3EW

Farm Animal Welfare Council
  Government Buildings, Hook Rise South, Tolworth,
  Surbiton, Surrey KT6 7NF

Humane Slaughter Association
  34 Blanche Lane, South Mimms, Potters Bar, Herts EN6 3PA

International Federation of Organic Agriculture Movement
(IFOAM)
  Okozentrum Imsbach, D-6695 Tholey-Theley, Germany

Meat and Livestock Commission
  Queensway House, Bletchley, Milton Keynes MK2 2EF

Milk Marketing Board
  Thames Ditton, KT7 0EL

National Dairy Centre
  5-7 John Princes Street, London W1M 0AP

Organic Farmers and Growers
  9 Station Approach, Needham Market, Stowmarket, Suffolk
  IP6 SAT

Rare Breeds Survival Trust
  National Agricultural Centre, Stoneleigh, Kenilworth,
  Warwickshire CV8 2LG

Smallholder Publications (*Smallholder* monthly magazine)
  High Street, Stoke Ferry, Kings Lynn, Norfolk PE33 9SF

Soil Association
  86 Colston Street, Bristol BS1 5BB

United Kingdom Register of Organic Food Standards
(UKROFS)
  Food From Britain, 301-344 Market Towers, New Covent
  Garden Market, Nine Elms Lane, London SW8 5NQ

Universities Federation for Animal Welfare (UFAW)
  8 Hamilton Close, South Mimms, Potters Bar, Herts
  EN6 3QD

Bibliography

Black, Maggie, *Home-made Butter, Cheese and Yoghurt* (E.P. Publishing, London, 1977)

Bruce, M.E., *Common-sense Compost Making* (Faber & Faber, London, 1946)

Chamberlain, A.T., Walsingham, J.M., and Stark, B.A. (eds), *Organic Meat Production in the 90s* (Chalcombe, Princes Risborough, 1989)

Cooper, Michael, *Discovering Farmhouse Cheese* (Shire, Princes Risborough, 1978)

de Bairacli Levy, Juliette, *Herbal Handbook for Farm and Stable* (Faber, London, revised 1973)

Dubach, Josef, *Traditional Cheesemaking* (translated Bill Hogan) (Intermediate Technology Publications, London, 1987)

Foster, L, 'Herbs in pastures: Development and Research in Britain, 1850-1984', *Biological Agriculture and Horticulture* 5(2):97-133 (1988)

Fraser, Andrew F., *Farm Animal Behaviour* (Bailliere Tindall, London, 1980)

Jacobs, Linda, *Environmentally Sound Small-scale Livestock Projects* (CODEL/Heifer Project International/VITA, 1986)

Kilgour, Ronald, and Dalton, Clive, *Livestock Behaviour* (Granada, London, 1984)

Lampkin, Nicolas, *Organic Farming* (Farming Press, Ipswich, 1990)

Macleod, G., *The Treatment of Cattle by Homoeopathy* (Health Science Press, Saffron Walden, 1981)

Porter, Valerie, *Animal Rescue* (Ashford, Shedfield, 1989) *Practical Rare Breeds* (Pelham, London, 1987)

Reynolds, Peter J., *Iron-Age Farm: The Butser Experiment* (British Museum Publications, London, 1979)

Sainsbury, David, *Farm Animal Welfare* (William Collins, London, 1986)

Scott, W.N. (ed), *The Care and Management of Farm Animals* (Bailliere Tindall, London, 1978)

Thear, Katie, *The Home Dairying Book* (Broad Leys Publishing, Saffron Walden, 1978)

Thear, David and Katie, *The Home Farm Sourcebook* (Broad Leys Publishing, Saffron Walden, 2nd edition 1990)

Toussaint Raven, E., *Cattle Footcare and Claw Trimming* (Farming Press, Ipswich, 1985)

Turner, Newman, *Fertility Farming* (Faber, London, 1951) *Fertility Pastures* (Faber, London, 1955)

Young, Rev. Arthur, *General View of the Agriculture of the County of Sussex* (1813)

Universities Federation for Animal Welfare:

Management and Welfare of Farm Animals (Bailliere Tindall, London, 3rd edition 1988)

Animal Training (proceedings of symposium, 1989)

Extensive and 'Organic' Livestock Systems: Animal Welfare Implications (proceedings of symposium, 1990)

Index